建筑立场系列丛书 No.71

C3

公共建筑
Architecture for the City

大都会建筑事务所 等 | 编

孙探春 于风军 杜丹 王京 时敏 林英玉 | 译

大连理工大学出版社

C3 建筑立场系列丛书 No.71

城市建筑：私人时代的公共建筑

详谈社区建筑设计

C3 No.71 Architecture for the City

Architecture for the City
Public Buildings in a Private Time

Narratives for Community Architecture

Architect for the Public B in a Priva

城市建筑：私人时代的公共建筑

西方社会对公共空间的理解主要与公共空间是一个民主大众的场所这一想法有关。此外，公共空间往往是自由和机会的代名词，个人和集体都可以在此畅所欲言，展现自我。希腊的集会广场或罗马的论坛可以说是公共空间最典型的代表。然而，当今这个时代，在日益充斥着虚拟演员的城市中，日常生活的动态对与公共空间概念相联系的文化品位和政治素质提出了质疑。城市建筑在这个过程中起到了重要的作用。作为集体记忆的宝库，城市建筑是给公共空间带来活力的根本媒介。事实上，像拱廊、大门、入户门、窗户或者阳台这些不受时间影响的建筑元素已经实现了公共生活与家庭领域之间的过渡。现在，这种过渡日益转变为通过玻璃幕墙、超大屏幕电视和智能手机来实现。

The western world's understanding of a public space is chiefly connected with the idea of a democratic place. Further, public space is often synonym of freedom and opportunity for individual expression and collective performance. The Greek agora or the Roman forum are arguably the most canonical figures of public space. However, in this day and age, the dynamics of everyday life in an urban world increasingly populated with virtual actors challenges the cultural and political qualities that we associate with the notion of public space. The architecture of the city plays an important role in this process. As a repository of collective memory, it is a fundamental agent to activate the public spaces. Indeed, the transitions between public life and the domestic realm have been chiefly negotiated through timeless architectural devices such as arcades, gates, doors, windows or balconies. Now, they are being increasingly negotiated by glazed facades, Jumbotron's, and smartphones.

皮埃尔·拉松德展览馆_Pierre Lassonde Pavilion/OMA

Dokk1图书馆_Dokk1/Schmidt Hammer Lassen Architects

葡萄酒之城博物馆_Cité du Vin/XTU Architects

斯塔夫罗斯·尼亚尔霍斯基金会文化中心

Stavros Niarchos Foundation Cultural Center/Renzo Piano Building Workshop

城市建筑：私人时代的公共建筑

Architecture for the City: Public Buildings in a Private Time/Nelson Mota

由意大利建筑师、测量师詹巴蒂斯塔·诺利于 1748 年绘制的著名罗马地图把建筑内部（如万神殿）和广场（如纳沃纳广场）作为公共空间展示出来。在这两个案例中，其空间都代表着欲望和冲突，体现了这个城市的人们在生活中的挣扎和矛盾。更近代一些，从第二次世界大战后期一直到 20 世纪 80 年代，受欧洲和北美洲国家福利政策的影响，建筑师们常常面临挑战：如何把民主的公民空间延伸至公共建筑内部。正如卡尔·波普尔所说的，这是开放社会这一理念的全盛时期。如今，建筑物——通常被视为纪念碑或者修辞表现手法——起着像广场、城市广场或城市公共场所等建筑类型的作用，作为社交互动和公共交往的主要场所。公共空间正在逐渐渗透到私人的壳体中。

The famous map of Rome drawn in 1748 by the Italian architect and surveyor Giambattista Nolli showed the interior of buildings such as the Pantheon and squares such as Piazza Navona as public spaces. In both cases, these were spaces of desire and conflict, places that expressed the struggles and contradictions of the life in the city. More recently, under the influence of the politics of the welfare state that ruled in Europe and North America from the post Second World War until the 1980s, architects were often challenged to expand the democratic civic space into the inside of public buildings. This was the heyday of the idea of open society, as Karl Popper put it. Now buildings – often seen as monuments or rhetorical devices of representation – are taking over the role of architectural types such as the piazza, the town square or the city commons as the main locus for social interaction and public representation. Public spaces are increasingly penetrating into private shells.

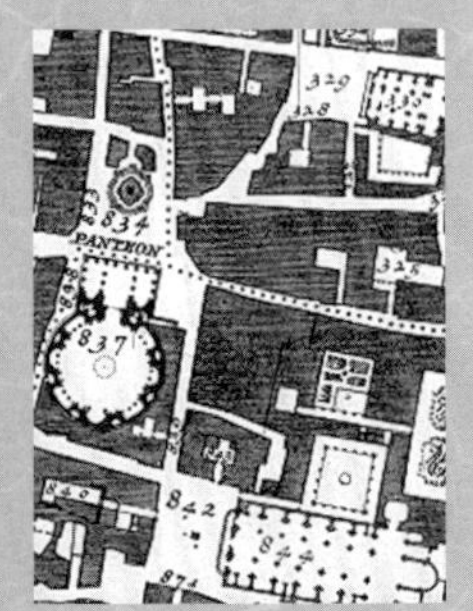

诺利地图，詹巴蒂斯塔·诺利，1748年
The Nolli map, Giambattista Nolli, 1748

城市建筑：私人时代的公共建筑

1990 年，美国建筑师和艺术家维托·阿肯锡在《批判性探索》（Critical Inquiry）杂志上发表了一篇文章，最后用一句高深莫测的话结尾："当心随身听"[1]。虽然看起来神秘难解，但是我们可以把它解读为阿肯锡给生活在电子时代的人们的一句忠告。在这个时代，公共这一概念只不过是由私有部分组合而成的。现在，在公共空间中，头戴式耳机司空见惯，这证明了一个事实：只有当我们可以安全地沉浸在个人世界中的时候，似乎才能够忍受处在如火车站、广场、市场、公园这样的公共建筑和公共空间中。智能手机保护着连接我们与我们的舒适区，即属于自己的世界的那根脐带。

理查德·桑内特在其名著《公共人的衰落》一书中描述了我们的公共生活逐渐私有化的过程。作为人们聚集的空间，公共空间一直在减少。例如，在建筑媒体上，我们读到的令人振奋的新广场或者新公园的设计数量远不及新颖的图书馆或出色的新博物馆。虽然在开放的公共空间中营造场所感越来越难，但是把人们聚集到封闭、独立的空间里进行社交互动却越来越容易。维托·阿肯锡认为，这是因为我们需要逃离充满政治色彩和我们无法控制的空间，需要在人群聚集的地方寻求庇护，因为在这种地方，我们能得到正在寻找的东西，有时得到的是正在为之付出代价的东西。

前面提到的项目都是新建筑的典范，它们将最终取代中规中矩的公共空间形式。博物馆、图书馆、歌剧院的定位恰到好处，将取代广场和公园成为历史古迹，成为弥漫着子孙后代的记忆和想象的地方。回顾这些公共建筑的设计可以帮助我们了解一些有利于这种趋势传播的因素。

Architecture for the City: Public Buildings in a Private Time

The American architect and artist Vito Acconci concluded an article published in 1990 in the magazine *Critical Inquiry* with the enigmatic sentence: "Beware of the Walkman"[1]. While apparently mysterious, this can be read as Acconci's piece of advice for a society living in an electronic age where the idea of public is nothing more than a combination of private parts. The widespread presence of headphones in the public space that we experience these days testifies to the fact that we seemingly tolerate being in public buildings and civic spaces such as train stations, squares, market halls, and public parks only when we can be safely immersed in our private world. The smartphone preserves the umbilical cord that connects us to our comfort zone, our domestic realm.

With "the Fall of Public Man", as Richard Sennet famously described the progressive process of privatization of our life in common, the public space has been losing territory as a space that gathers. In architectural media, for example, we read much less about a new and inspiring design for a square or a public park than about an original library or an outstanding new museum. While it has become increasingly difficult to create a sense of place in public open spaces, it is gradually easier to gather people in closed, contained spaces for social interaction. Vito Acconci suggests that this is a consequence of our need to run away from spaces that are politically charged and somehow out of our control, and get refuge in cluster places where we can get what we are looking for, and sometimes paying for.

The projects featured ahead provide excellent examples of new buildings that will eventually replace the canonical types of public space. The museum, the library, the opera are well positioned to replace the squares and parks as the historical places that will pervade the memory and imagination of future generations. Reflecting on the design of these public buildings can help us understand some of the factors that contribute for the dissemination of this trend.

斯塔夫罗斯·尼亚尔霍斯基金会文化中心，Renzo Piano Building Workshop
Stavros Niarchos Foundation Cultural Center, Renzo Piano Building Workshop

葡萄酒之城博物馆，XTU Architects
Cite du Vin, XTU Architects

作为"人造地形"的公共建筑

把建筑设计得如同地貌特征似乎暗示了设计师设计风格的质朴，故意抹去了设计师在改变建筑环境中的作用。然而，在大多数情况下，事实恰恰相反。这样的建筑设计极其复杂，需要高水平的专业知识，是在设计上大胆尝试的明显例证。伦佐·皮亚诺设计的斯塔夫罗斯·尼亚尔霍斯基金会文化中心便是如此。皮亚诺和他的团队创造了一个奇迹：将一个包括希腊国家图书馆和希腊国家歌剧院等其他设施在内的非常复杂的建筑，天衣无缝地融入到170 000m² 的景观公园中。有许多项目，其十分复杂、要求极高的设计与建筑物的简洁朴素形成了鲜明的对比。斯塔夫罗斯·尼亚尔霍斯基金会文化中心只是其中一个。实际上，如果从雅典卫城来到这一文化中心所在的雅典卡利地亚区，看到和一公顷足球场一样大的悬浮屋顶悬浮在坡度平缓的绿色公园之上时，你可能会感到惊奇。

这座建筑的实际大小仅能从码头或者沿岸繁忙的街道才能看出来。这一建筑有着重要的政治立场。在像巴黎或维也纳这样的城市里，歌剧院和国家图书馆具有纪念意义，它们强调的是一个分界线，将那些有幸能进入这些城市殿堂的少数人与所有那些无法进入其中的人分离开来。在雅典，也许受到这所城市在建立民主方面所起的重要作用的启发，皮亚诺和他的团队有意用绿色植物立面和玻璃幕墙将建筑遮掩起来，吸引路经此处的无名小卒进去体验这一文化中心，把它当成公共空间真正意义上的延伸。

作为展示工具的公共建筑

如果说在雅典，建筑设计方法刻意追求建筑和开放空间的融合，

Public Buildings as Invented Topographies

Designing a building as a topographical feature seemingly suggests an act of designer's modesty. A deliberate attempt to obliterate the designer's agency in the transformation of the built environment. In most cases, however, the opposite is true. It is an exercise of extreme complexity and demands a high level of expertise. It is a clear example of design bravura. This is the case of Renzo Piano's project for the Stavros Niarchos Foundation Cultural Center. Piano and his team made the prodigy of hiding an incredibly complex program that includes, among other facilities, the National Library of Greece and the Greek National Opera seamlessly integrated in a landscaped park with 170,000m². This is one of those projects where there is a striking contrast between a very complex and demanding program and the simplicity of the building that accommodates such complexity. Indeed, someone coming from the Acropolis through Athens' Kallithea district, where the cultural center is located, will probably be surprised by the presence of a floating roof with the same area as a football field – one hectare – hovering over a green park with a gentle slope.

The real scale of the building is only perceived from the side of the marina and the busy roads along the shore. There is an important political statement in this architectural operation. The monumentality of the opera houses and national libraries in cities such as Paris or Vienna emphasizes the boundary that separates the few lucky ones that can get inside these temples of the bourgeois city and all the others that cannot do the same. In Athens, maybe inspired by the city's seminal role in the foundation of democracy, Piano and his team deliberately created a sort of camouflage made out of greenery and glazed facades that invites the anonymous passer-by to go inside and experience the cultural center as a true extension of the public space.

Public Buildings as Instruments of Display

If in Athens the architectural approach deliberately pursued a fusion between built mass and open space, in the "Cité

皮埃尔·拉松德展览馆，OMA
Pierre Lassonde Pavilion, OMA

那么 XTU 建筑事务所在波尔多设计的"葡萄酒之城"博物馆（"Cité du Vin"）称颂的则是独特的建筑形式和材料，唤起人们对当地葡萄酒传统的记忆。建筑师直接探索利用独特的几何外形来彰显建筑在周围景观中的存在。建筑的内部和外部都没有设计得模棱两可的地方。建筑物顶礼膜拜的上好葡萄酒也不可能没有特色。相反，它们需要刺激感官，激发情感。同样，"葡萄酒之城"推出的感官体验与人们平凡的日常生活形成了对比。就像波尔多上好的葡萄酒一样，"葡萄酒之城"是一座令人身心迷醉、精心设计的建筑，能够激起人们无尽的感官体验。

与"葡萄酒之城"的感性外形形成鲜明对比，大都会建筑事务所设计的魁北克国立美术馆新馆皮埃尔·拉松德展览馆是一座"不那么张扬，甚至有点儿隐秘之感的城市新建筑"，正如这家荷兰事务所自身所言。事务所采取循环反复的建筑设计方法，将这个置身于公园之中的展馆设计成三个堆放的体量，从地面层开始逐渐向上至顶层，大小依次递减。在顶层，游客可以饱览魁北克市的风光。一段螺旋楼梯吸引游客进入一条长廊，长廊连通两个安装了玻璃幕墙的下层体量。在二层，长廊突然通到外面，变成一个向上倾斜的体量，连接上面二层画廊。这条观光路线似乎与美术馆硬朗的几何外形正相反，成了促进该建筑与周围环境相融合的序曲。事实上，正如大都会建筑事务所所说的："该建筑旨在把城市、公园与博物馆相互联系起来，同时成为这三者的延伸。"公共空间渗透到建筑物中，反之，建筑物也渗透到了公共空间中。

建筑师希望建筑能扩大公共空间传统界限的雄心抱负也体现在 Dokk1 图书馆这一项目上。Dokk1 图书馆是丹麦最大的公共图书馆，由丹麦的施密特－哈默－拉森建筑事务所设计。Dokk1 图书馆是一个巨大的多边形结构，占据了奥胡斯河口的绝佳位置。利用事务所在设计

du Vin" designed by XTU Architects in Bordeaux, there is a celebration of distinctive forms and materials to evoke the region's wine tradition. The architects bluntly explore the strangeness of its geometric composition as an asset to reveal the building's presence in the landscape. Both inside and outside the building there is no room for neutrality. Good wines, which are the products that the building pays homage to, cannot be neutral either. Rather, they need to activate the senses and stimulate emotions. Likewise, the sensual experience promoted by "Cité du Vin" competes with the ordinariness of people's everyday life. Like the good wines of Bordeaux, the "Cité du Vin" is a voluptuous and carefully crafted building that can nevertheless provoke overstimulation of the senses.

Contrasting to the sensual form of "Cité du Vin", OMA's Pierre Lassonde Pavilion of the Musée National des Beaux-arts du Québec is a "subtly ambitious, even stealthy, addition to the city", as the Dutch office itself recognizes. Resorting to a recurrent formula in the office's recent work, the pavilion is structured in three stacked volumes decreasing in size from the park level surrounding the museum to the top where the visitors can enjoy a broad perspective of Quebec City. A curvy stair invites the visitors to engage in a promenade that connects the two lower glazed volumes and then bursts onto the outside as a tilted volume to connect the two upper gallery boxes. This scenic route seems to offer a counter form to the rigid geometry of the gallery and performs as a teaser to stimulate the merging of the building with its surroundings. Indeed, as OMA states, "the building aims to weave together the city, the park and the museum as an extension of all three simultaneously." The public space penetrates the building and the other way round.

The ambition of creating buildings that expand the traditional limits of the public space can also be seen in the project for Dokk1, the largest public library of Denmark, designed by the Danish office Schmidt Hammer Lassen Architects. Dokk1 is a large polygonal volume that occupies a privileged location at the mouth of the Aarhus River. Dwelling on the office's exper-

Dokk1图书馆，Schmidt Hammer Lassen Architects
Dokk1, Schmidt Hammer Lassen Architects

世界上大量重要的新生代图书馆方面的专长，Dokk1 图书馆超越了图书馆作为文化机构的功能，成为城市生活不可缺少的一部分，完全消除了建筑内部与外部的界限。城市和建筑之间的界限也变得模糊。的确，事务所声称，该建筑本质上是带有屋顶的城市空间，促进知识交流，增加各种机遇。它也是一个民主的空间，供民众参与，有意识地避免层级的划分。该建筑没有主立面或背面，足以证明设计师的这种想法。施密特－哈默－拉森建筑事务所在设计 Dokk1 图书馆时探索了如何将一个普普通通的城市基础设施与其内部空间巧妙融合。该建筑内部充满了斯堪的纳维亚氛围，鼓励人们进行适度的社会交往。

当心智能手机

以上讨论的四座公共建筑很好地证明了维托·阿肯锡在 20 世纪 90 年代初的预言。他写道："在电子时代，公共空间是移动的空间。你匆匆来访，不作停留。"实际上，正是因为人们在公共空间只作短暂停留才促使设计师在设计像图书馆、歌剧院或博物馆这样的公共建筑时采用了新的设计方法。"希腊的集会广场"现在只是像斯塔夫罗斯·尼亚尔霍斯基金会文化中心这样的建筑综合体的一部分。"罗马的论坛"也在像 Dokk1 图书馆这样的建筑中改头换面。但这些新的集会广场和论坛都有开放时间、保安和监控摄像头。此外，当我们身处这些新的公共空间时，往往感觉它们就是私人世界的一部分，耳机保护着我们，并把我们与其他空间，譬如虚拟空间，连接起来。如书中前面所提到的那些新一代公共建筑，它们使我们对汇集了个人感受的公共空间有了新的理解。用维托·阿肯锡的话说，这些都是私人时代的公共建筑。当心智能手机。

tise in designing a great number of important new generation libraries around the world, Dokk1 goes beyond the role of the library as a cultural institution. It makes it part and parcel of city life and stimulates a permanent negotiation of the limits that divide the interior of the building with its exterior. The limits between the city and the building are blurred. To be sure, the office claims that the building is essentially a covered urban space that promotes the exchange of knowledge and opportunities. It is also a democratic space, open to civic participation, where there is a deliberate attempt to avoid hierarchical compositions. The lack of a main or back facade testifies to this ambition. In the design of Dokk1, Schmidt Hammer Lassen Architects explores a well-crafted fusion of a rough piece of urban infrastructure with interior spaces where there is a Scandinavian atmosphere that stimulates moderated social interactions.

Beware of the Smartphone

The four public buildings discussed above are good examples of Vito Acconci's prophetic words in the early 1990s: "Public space, in an electronic age, is space on the run," he stated. "You come to visit, not to stay". It is indeed this transient experience of the public space that underpins the emergence of new approaches in the design of civic buildings such as public libraries, opera houses or museums. The Greek agora is now part of building complexes like the Stavros Niarchos Foundation Cultural Center. The Roman forum has been reconceptualized in buildings like Dokk1. But these new agorae and forums have opening hours, security guards and surveillance cameras. Furthermore, we often experience these new public spaces as part of our private world, protected by headphones that connect us to other spaces, virtual spaces. A new generation of public buildings like those featured in the pages ahead caters for a new understanding of the public space as an assemblage of individual experiences. Paraphrasing Vito Acconci, these are public buildings for a private time. Beware of the smartphone. Nelson Mota

1. Vito Acconci, "Public Space in a Private Time", *Critical Inquiry*, Vol 16, no. 4, 1990, p.900~918

皮埃尔·拉松德展览馆

OMA

皮埃尔·拉松德展览馆是魁北克国立美术馆(MNBAQ)的第四座建筑，既与老馆彼此连通，又自成一体。作为一栋新的城市建筑，其设计不是那么野心勃勃，甚至有点儿隐秘之感。皮埃尔·拉松德展览馆不是强加于这个城市的一个图标，而是形成了公园和城市之间的新连接，增强了魁北克国立美术馆的凝聚力。

新展馆周围的环境复杂而敏感，使设计需要考虑如下关键问题：如何延伸国家战场公园而又与城市结合？如何在尊重和保护圣多米尼克教堂的同时在格朗达莱大街建造一座令人心悦诚服的建筑？如何使美术馆的组织结构明晰同时又扩大规模？大都会建筑事务所的解决方案是把新展馆建成三个相互叠加、大小依次递减的体量，里面包括：临时展区、现代和当代藏品永久展区、装饰艺术和设计展区以及因努伊特艺术品展区。三个体量层叠在一起，如同瀑布，从城市流向公园。该建筑旨在使城市、公园和博物馆交织为一体，同时成为这三者的延伸。该建筑的三个体量呈台阶式逐层降低，每一层在平面上都向前伸出一节，整体将原来的教堂庭院框在里面，使整个建筑朝向公园。通过天窗和精心巧妙设计的窗户，公园景色融入了博物馆；通过将展览延伸至屋顶露台和凸出的室外楼梯，博物馆也融于公园之中。这种层叠的设计营造了一个12.6m高的大厅，位于壮观的长达20m的巨大悬臂之下。大厅作为博物馆通往格朗达莱大街的入口，也是一个实现博物馆公共职能的城市广场，这儿还是通向美术馆、庭院和礼堂的通道。与安静的、供人沉思冥想的展览空间形成互补，像门厅、休息室、商店、连桥、花园等功能区域都设计在博物馆的四周，人们既可以从事各种活动，也可以欣赏艺术品，同时也可漫步其间。一路上，游客可以从巨大的螺旋楼梯和凸出的室外楼梯上欣赏到精心设计的景致，这些景致使游客与公园、城市和博物馆的其他部分重新融为一体。在建筑内部，夹层和观景平台与临时展览空间和永久性展览空间相连。而在每个画廊体量的上面，其屋顶平台都为户外展览和活动提供了空间。新展览馆通过一条130m长的通道与博物馆原有的建筑相连，为博物馆收藏的、长达40m的保罗·里奥皮勒的画作《致罗莎卢森堡》营造了一个永恒的家。通过纯粹的长度和高度方面的变化，这条通道创造了一个令人惊奇的展览空间，好像不经意间把游客引向这一博物馆综合体的其他展馆。

悬臂结构由一个混合式钢桁架系统支撑，里面没有柱子，使展览空间连续而不间断。层次分明的外立面同时解决了结构、保温和采光三个方面的问题，满足了建筑物自然采光和应对魁北克冬季严寒的保温问题这一看似矛盾的需求。建筑立面采用了三层玻璃幕墙，由模仿桁架结构图案的2D烤漆玻璃、3D浮雕玻璃和漫射玻璃组成。在展馆

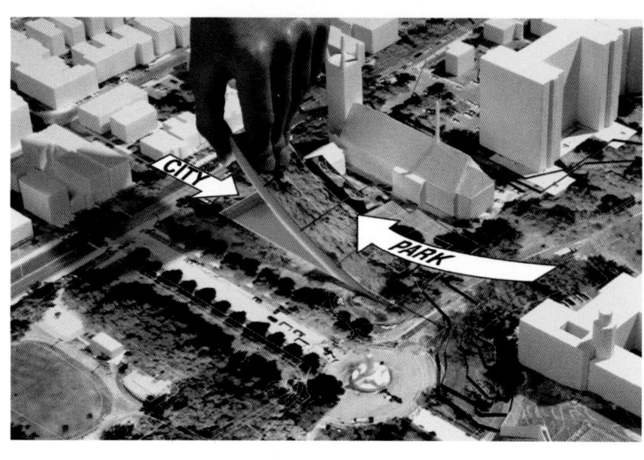

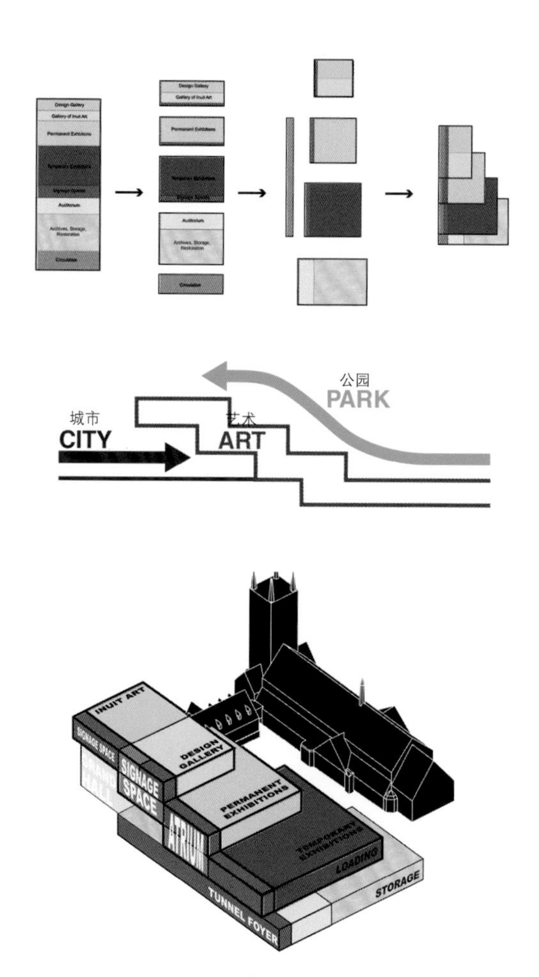

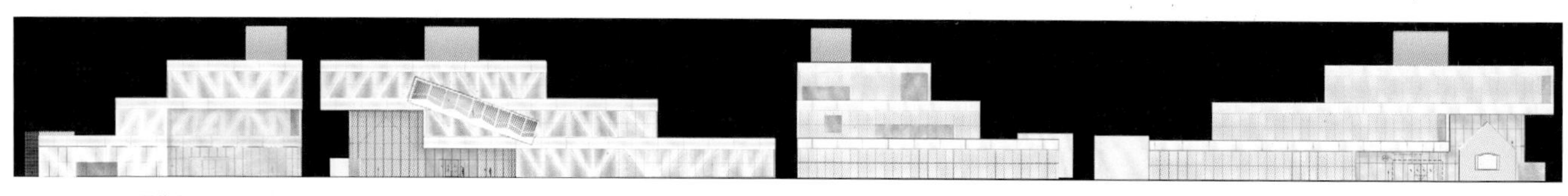

西北立面 north-west elevation　西南立面 south-west elevation　东南立面 south-east elevation　东北立面 north-east elevation

中，保温墙位于半透明玻璃的后面，两者之间有间隙，里面安装有照明设施。当这些灯光在晚上被点亮的时候，该建筑就像公园里的一盏灯笼。大厅的四面完全由玻璃幕墙包围。透过玻璃幕墙看过去，Charles Baillairgé 馆的美景几乎一览无余。半透明的展馆与清澈明亮的大厅之间鲜明的对比可帮助人们更好地解读该建筑的堆叠和悬臂体量。

Pierre Lassonde Pavilion

The Pierre Lassonde Pavilion – the Musée national des beaux-arts du Québec(MNBAQ)'s fourth building is interconnected yet disparate – is a subtly ambitious, even stealthy, addition to the city. Rather than creating an iconic imposition, it forms new links between the park and the city, and brings coherence to the MNBAQ.

The intricate and sensitive context of the new building generated the central questions underpinning the design: How to extend Parc des Champs-de-Bataille while inviting the city in? How to respect and preserve Saint Dominique church while creating a persuasive presence on Grande Allée? How to clarify the museum's organization while simultaneously adding to its scale? OMA's solution was to stack the required new galleries in three volumes of decreasing size to house temporary exhibitions, permanent modern and contemporary collections, and Decorative Arts and design, as well as Inuit artworks, creating a cascade ascending from the park towards the city. The building aims to weave together the city, the park and the museum as an extension of all three simultaneously. While they step down in section, the gallery boxes step out in plan, framing the existing courtyard of the church cloister and orienting the building towards the park. The park spills into the museum (through skylights and carefully curated windows) and the museum into the park (through the extension of exhibitions to the terraces and the outdoor pop-

项目名称：Expansion of the Musée national des beaux-arts du Québec (MNBAQ) / 地点：Parc des Champs-de-Bataille, Québec City, Canada / 建筑师：OMA
负责合伙人：Shohei Shigematsu / 项目团队：Jason Long, Ceren Bingol, Patrick Hobgood, Luke Willis, Rami Abou-Khalil, Richard Sharam, Tsuyoshi Nakamoto, Sandy Yum, Sara Ines Ruas, Ted Lin, Markus von Dellingshausen, Andy Westner, Jackie Woon Bae, Carly Dean with Sue Lettieri, Michael Jefferson, Mathieu Lemieux Blanchard, Martin Raub, Demar Jones, Cass Nakashima, Rachel Robinson
合作建筑师：Provencher_Roy Architectes (Montreal) / 结构：SNC Lavalin / 机电设计：Bouthillette Parizeau, Teknika HBA / 代码：Technorm / 声学：Legault & Davidson / 垂直交通：Exim / 造价控制：CHP Inc. / 照明设计：Buro Happold / 立面设计：FRONT / 立面工程：Patenaude Trempe, Inc., Albert Eskenazi, CPA structural Glass / 礼堂设计：Trizart Alliance / 当地咨询（竞赛）：Luc Lévesque / 承包商：EBC / 客户：Musée national des beaux-arts du Québec
建筑面积：14,900m^2_museum expansion (contemporary exhibitions: 5.5m high, 1,294m^2; permanent contemporary collection: 5m high, 912m^2; design and inuit galleries: 5m high, 535m^2; grand hall: 5.5m high_lower part, 12.6m high_higher part, 831m^2; grand stair: 79 steps_3 pieces, 15m long, 37 elements of curved glass; tunnel: 130.6m long (5.1m in elevation change); auditorium (no of seats): 256m^2; boutique: 263m^2; cafe: 140m^2; green roof: 3,327m^2 with 90,000 plants, 5 kinds of succulents; courtyard: 460m^2) / 高度：21.8m (26.5m including the cube) 4 floors (3 above grade) / 大厅悬臂结构：20m long, 12.5 m tall (floor to ceiling) / 材料：8,500m^3 of concrete, 1,090.000kg of steel, 1,195 panels of exterior glass, Three kinds of glass cover 95% of the exterior (53.3% opaque, 26.15% transparent, 20.35% translucent), glass panel sizes MR2_1,640mm x 2,530mm, glass panel sizes MR3_1,640mm x 3,330mm, glass panel sizes MR6_1,650mm x 1,965mm / 网格间距：Varies 3,500mm, 4,000mm, 5,000mm, 6,000mm
设计竞赛时间：2010 / 竣工时间：2016.6
摄影师：©Bruce Damonte (courtesy of the architect) - p.10~11, p.16~17, p.20~21, p.23[bottom], p.24~27, p.30~31; ©Nic Lehoux (courtesy of the architect) - p.28~29; ©Philippe Ruault (courtesy of the architect) - p.14~15, p.23[top];

LASSONDE

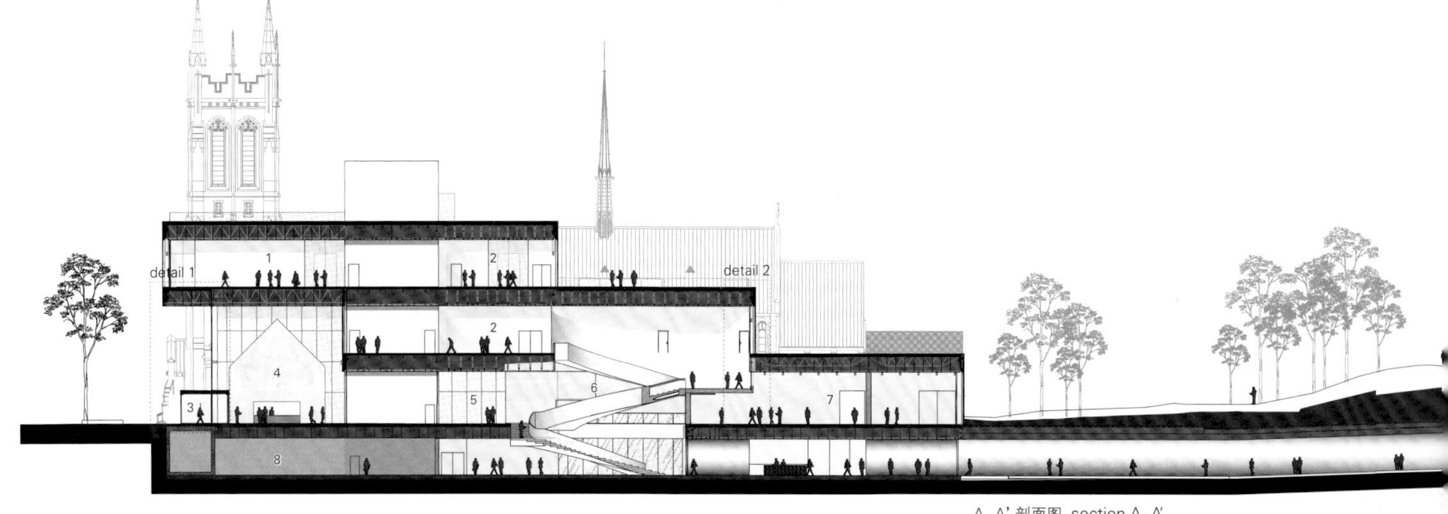

A-A' 剖面图 section A-A'

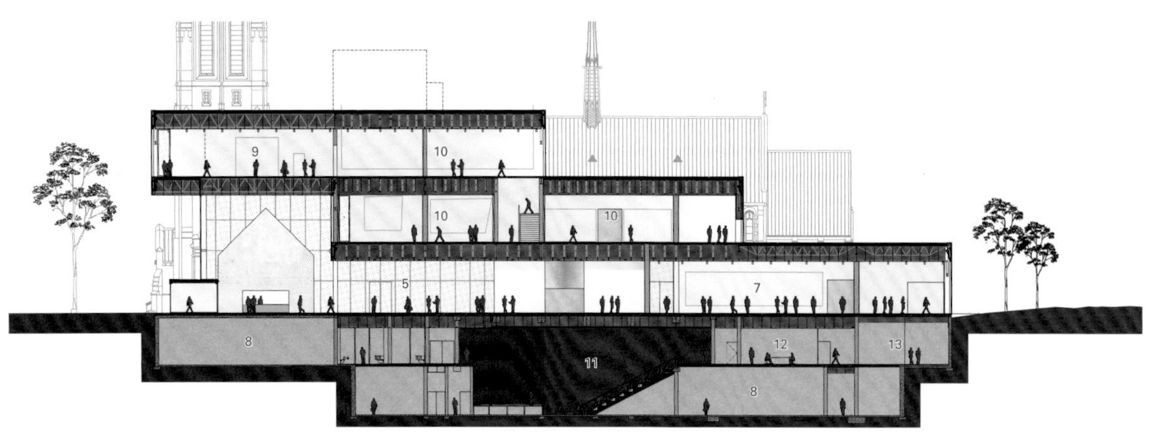

B-B' 剖面图 section B-B'

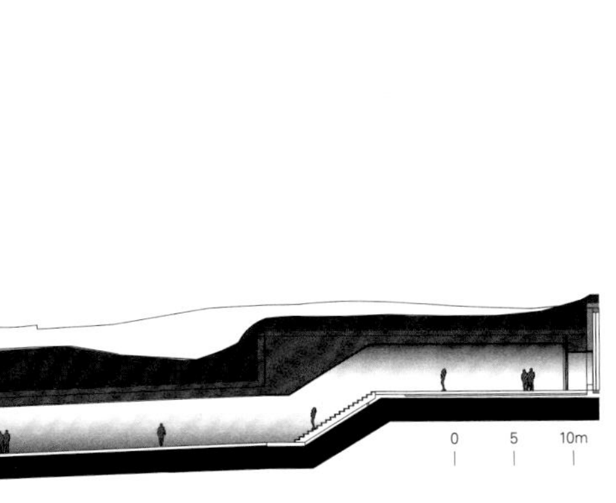

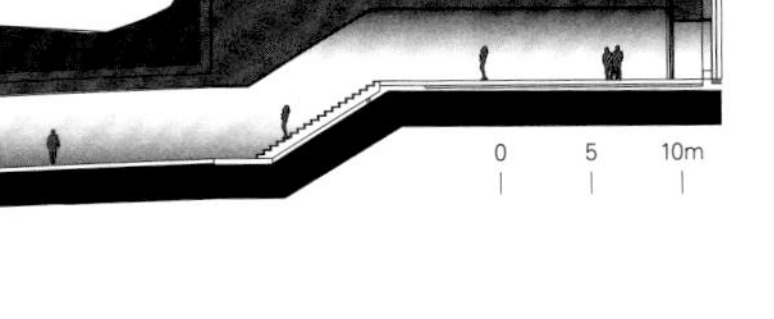

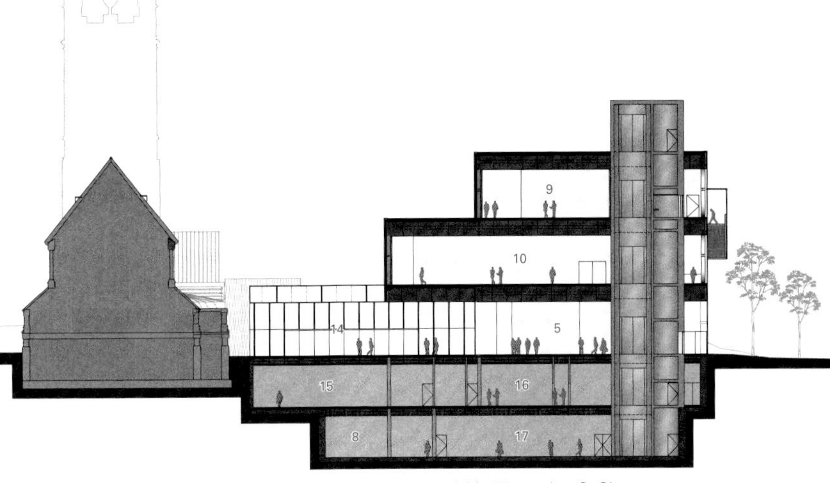

C-C' 剖面图 section C-C'

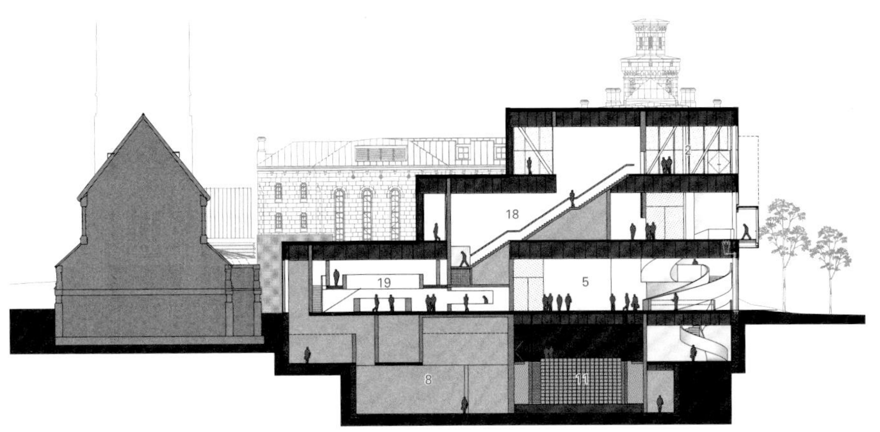

D-D' 剖面图 section D-D'

1. 多功能空间
2. 引导标示空间
3. 博物馆广场
4. 售票处
5. 大厅
6. 中庭楼梯
7. 临时展厅
8. 机械设备间
9. 因努伊特艺术品展区
10. 永久藏品展区
11. 礼堂
12. 办公室
13. 商店
14. 庭院
15. 板条箱储藏室
16. 浴室
17. 厨房
18. 疏散楼梯
19. 精品店

1. multi-purpose space
2. signage space
3. museum plaza
4. ticketing
5. grand hall
6. atrium stair
7. temporary gallery
8. mechanical room
9. inuit gallery
10. permanent collection
11. auditorium
12. office
13. shop
14. courtyard
15. crate storage
16. bathrooms
17. catering kitchen
18. egress stair
19. boutique

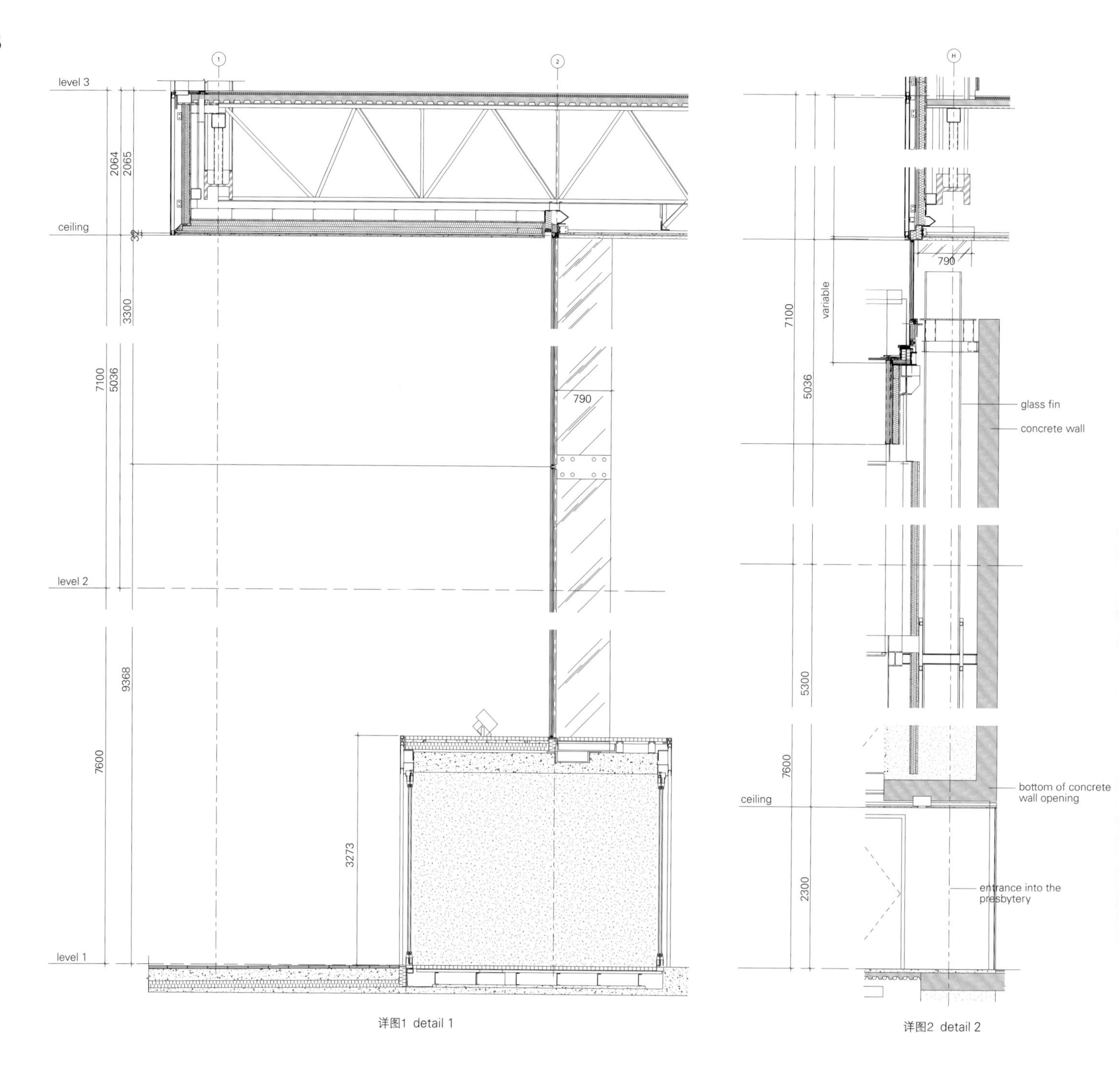

详图1 detail 1

详图2 detail 2

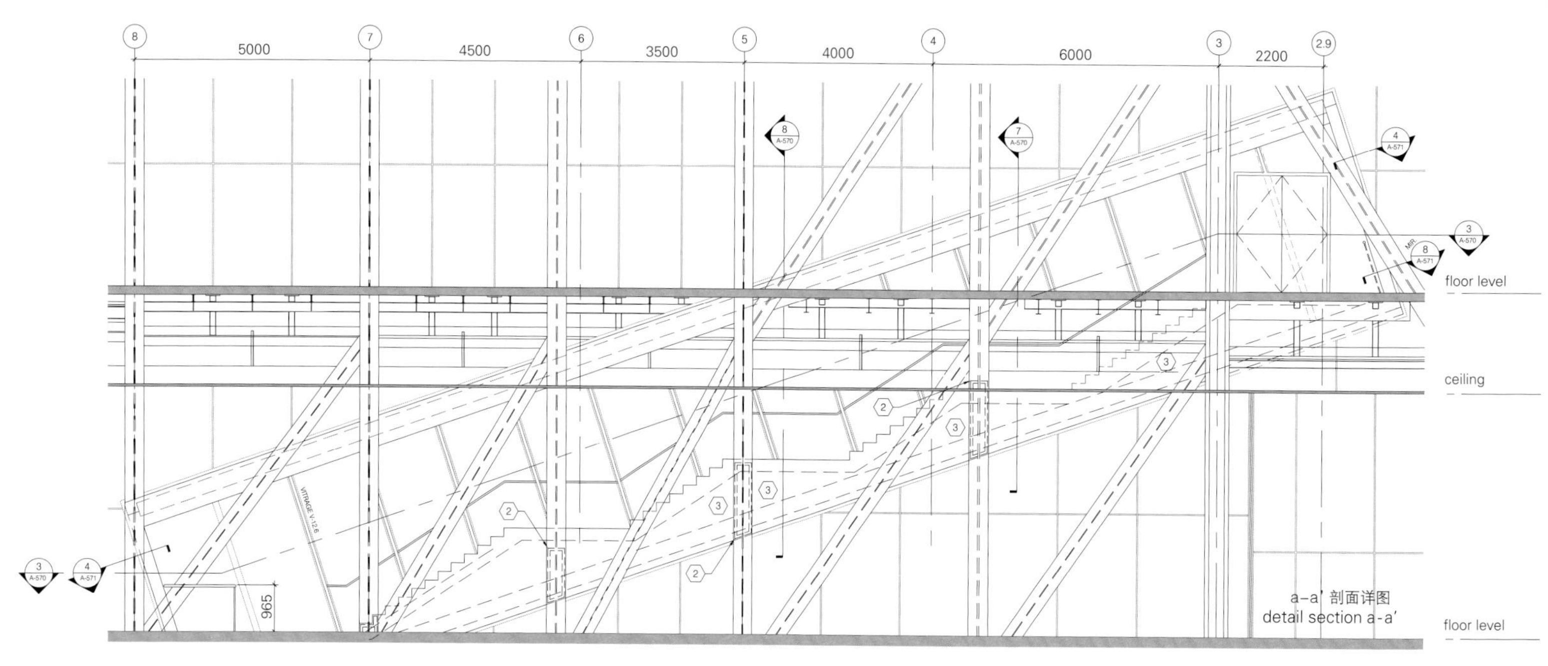

a-a' 剖面详图
detail section a-a'

三层半透明玻璃立面
triple glazed translucent facades

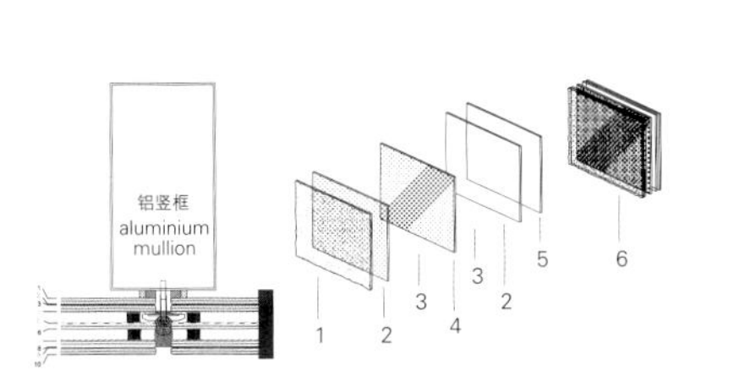

1. 5mm烧结花纹玻璃（统一的点状烧结块） 2. 6mm低辐射涂层玻璃 3. 12mm氩气填充
4. 6mm烧结玻璃（第二层带图形的烧结块） 5. 6mm透明玻璃 6. 三层中空玻璃
1. 5mm fritted & patterned glass (uniform dot frit) 2. 6mm low-e coated glass
3. 12mm argon 4. 6mm fritted glass (2nd layer of figurative frit) 5. 6mm clear glass 6. triple IGU

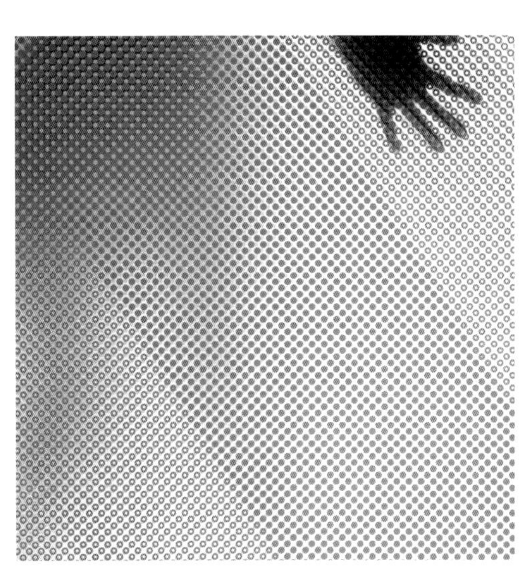

拱肩镶板和实心面板
spandrel and solid panels

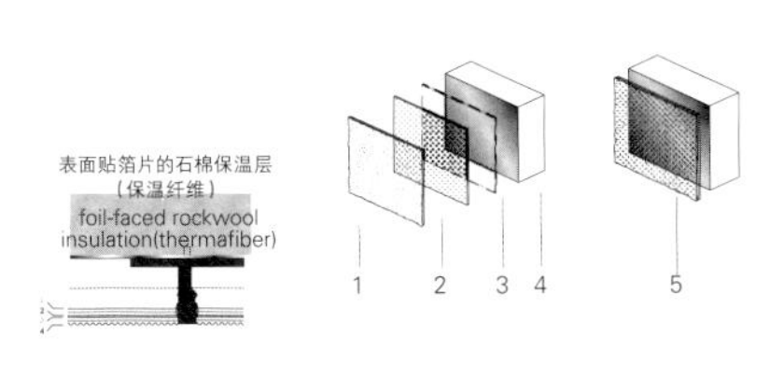

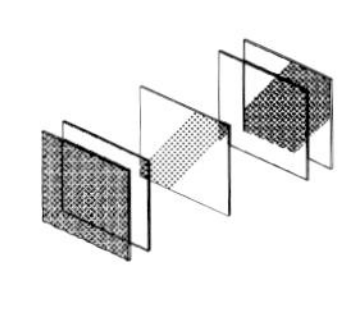

1. 5mm烧结花纹玻璃 2. 6mm烧结玻璃（统一的点状烧结块） 3. 穿孔金属板
4. 表面贴箔片的保温层（保温纤维或其他材料） 5. 实心面板
1. 5mm fritted & patterned glass 2. 6mm fritted glass(uniform dot frit)
3. perforated metal 4. foil-faced insulation (thermafiber or others) 5. solid panel

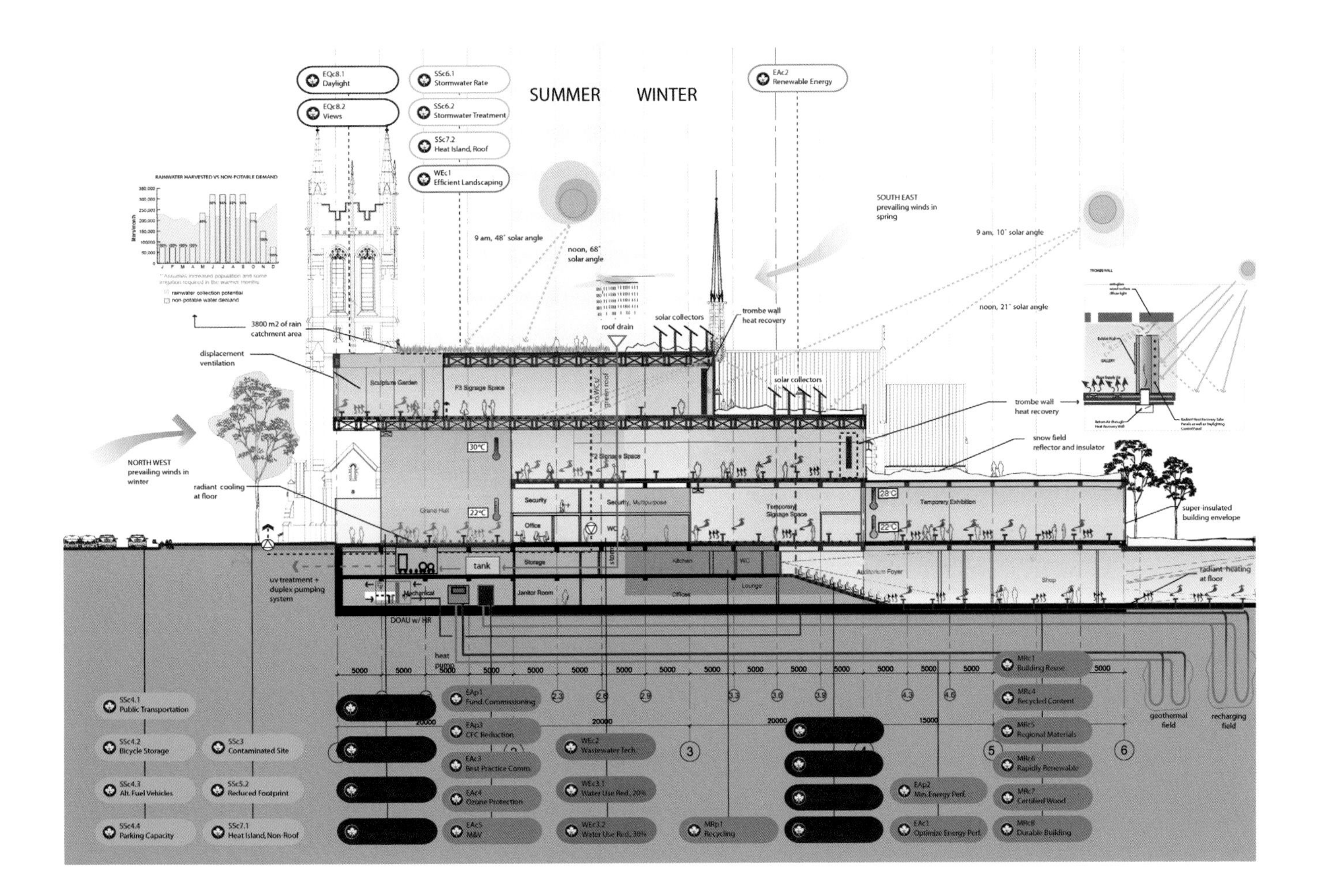

Excelsior

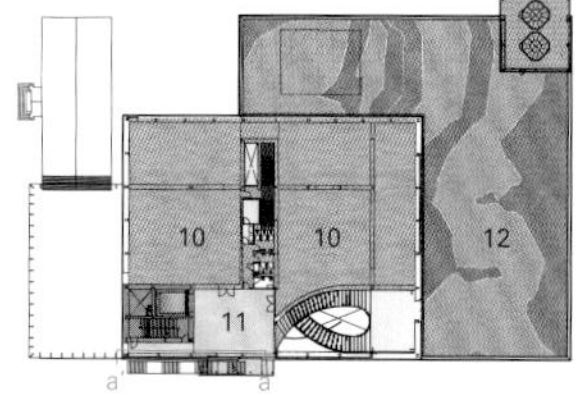

二层 second floor

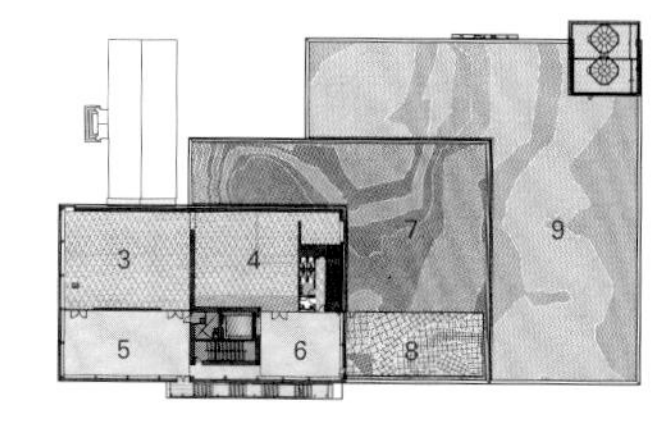

三层 third floor

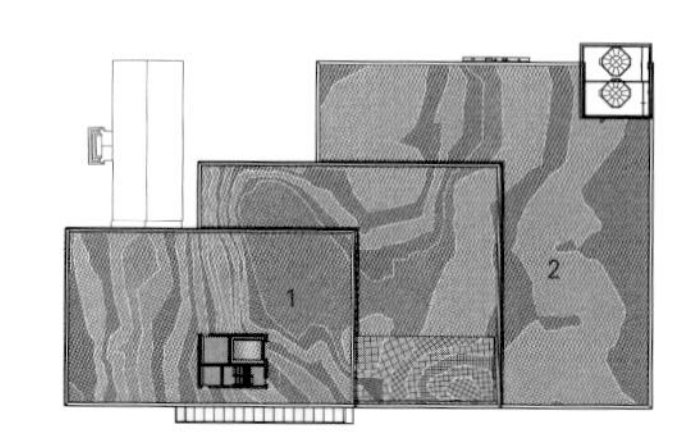

屋顶 roof

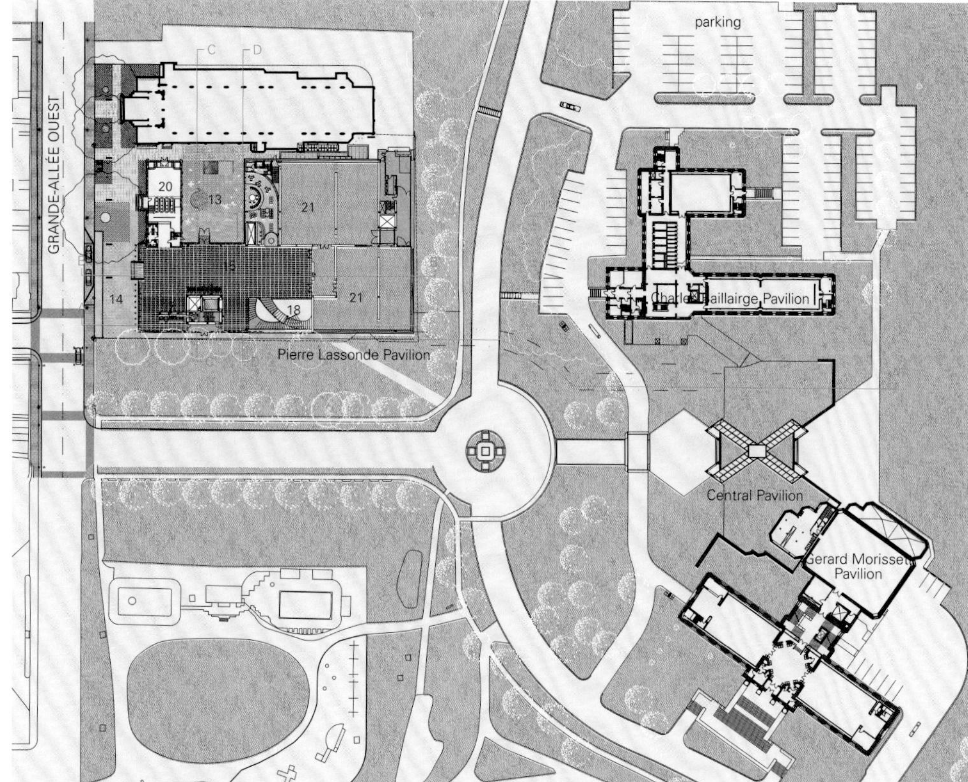

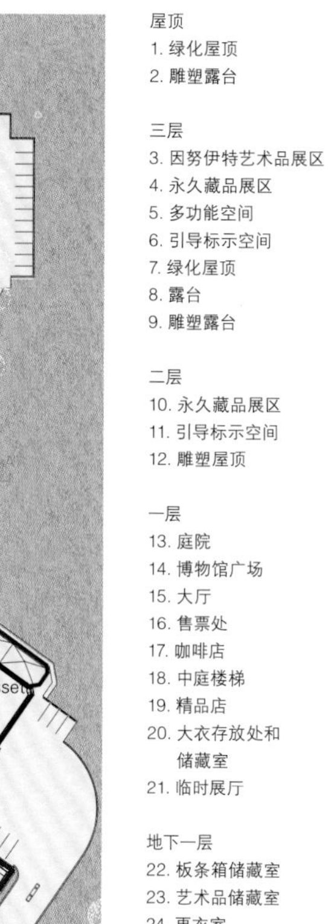

一层 ground floor

屋顶
1. 绿化屋顶
2. 雕塑露台

roof
1. green roof
2. sculpture terrace

三层
3. 因努伊特艺术品展区
4. 永久藏品展区
5. 多功能空间
6. 引导标示空间
7. 绿化屋顶
8. 露台
9. 雕塑露台

third floor
3. Inuit gallery
4. permanent collection
5. multi-purpose space
6. signage space
7. green roof
8. terrace
9. sculpture terrace

二层
10. 永久藏品展区
11. 引导标示空间
12. 雕塑屋顶

second floor
10. permanent collection
11. signage space
12. sculpture roof

一层
13. 庭院
14. 博物馆广场
15. 大厅
16. 售票处
17. 咖啡店
18. 中庭楼梯
19. 精品店
20. 大衣存放处和储藏室
21. 临时展厅

ground floor
13. courtyard
14. museum plaza
15. grand hall
16. ticketing
17. cafe
18. atrium stair
19. boutique
20. presbytery coat check and storage
21. temporary gallery

地下一层
22. 板条箱储藏室
23. 艺术品储藏室
24. 更衣室
25. 机械设备间
26. 休息室
27. 礼堂
28. 运送中的艺术作品
29. 办公室
30. 商店
31. 酒吧
32. Riopelle通道
33. 隧道

basement floor
22. crate storage
23. art storage
24. lockers
25. mechanical room
26. lounge
27. auditorium
28. works in transit
29. offices
30. shop
31. bar
32. Riopelle passage
33. tunnel

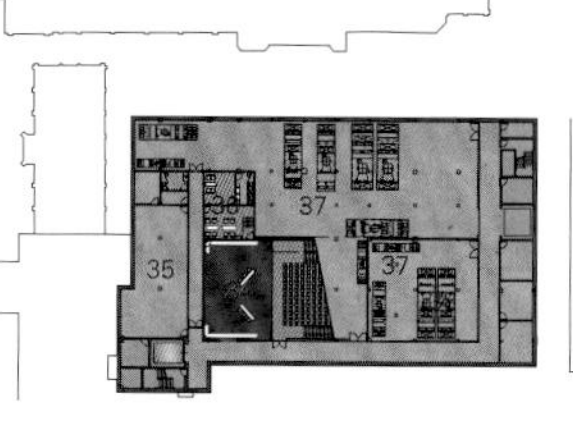

地下二层 sub basement floor

地下二层
34. 礼堂
35. 厨房
36. 暖房
37. 机械设备间
38. 办公室

sub basement floor
34. auditorium
35. catering kitchen
36. green room
37. mechanical room
38. offices

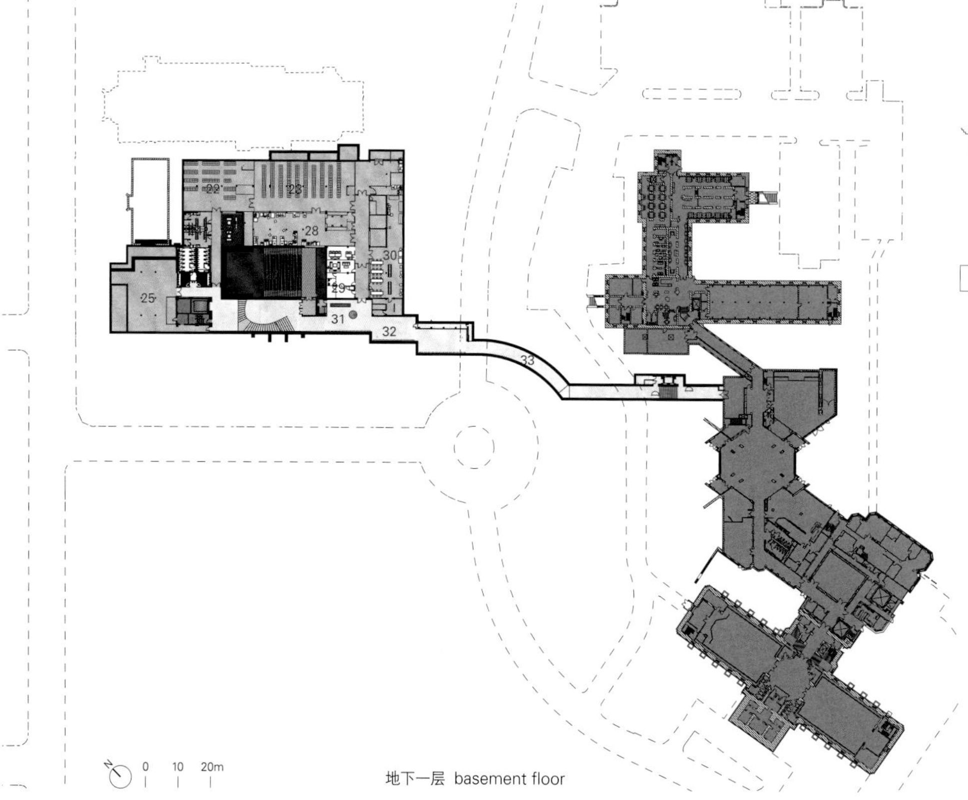

地下一层 basement floor

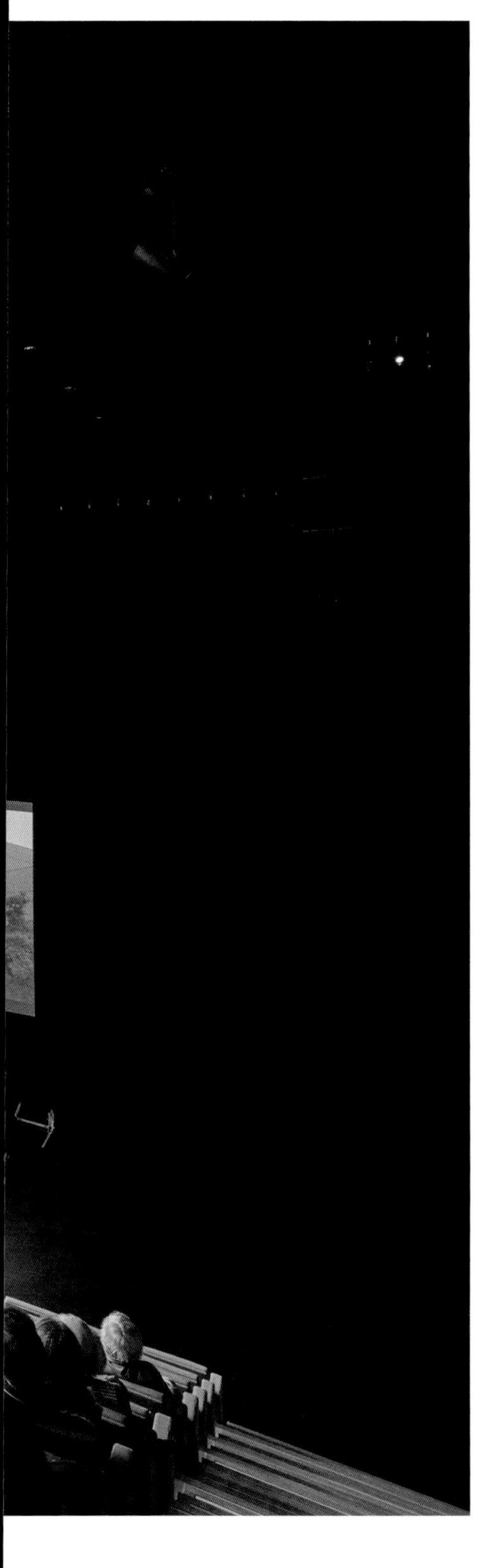

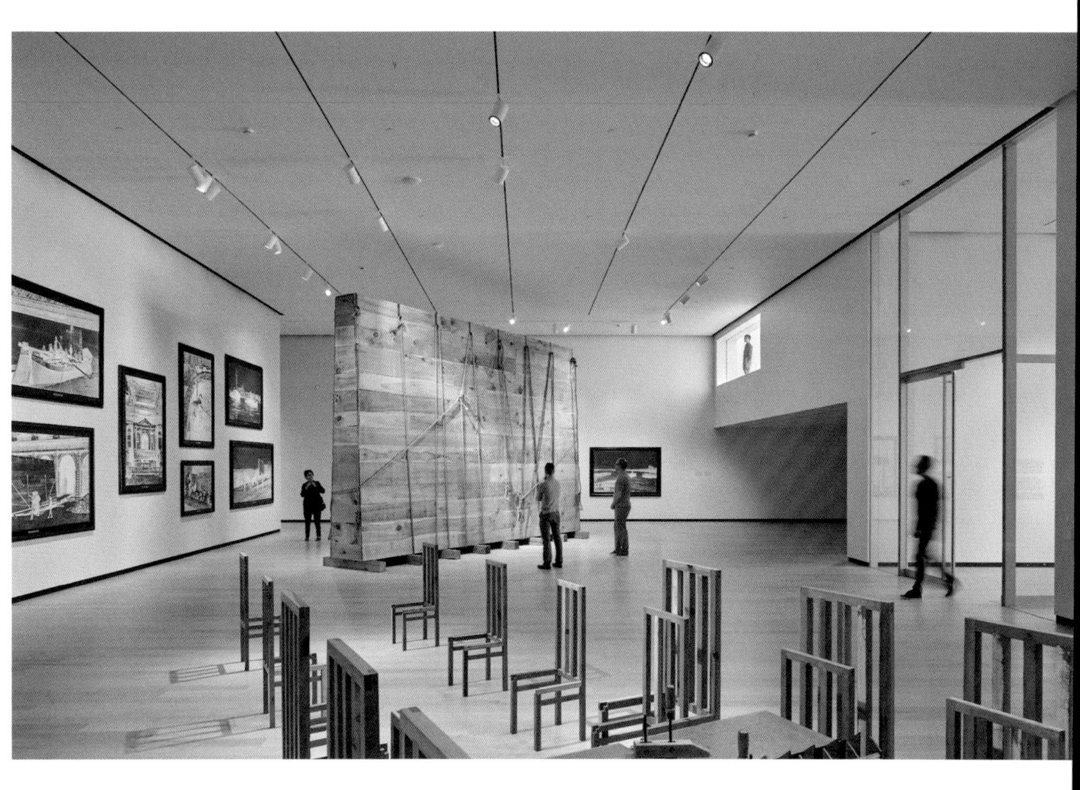

out staircase). The stacking creates a 12.6m-high (42 ft) Grand Hall, sheltered under a dramatic 20m (66 ft) large cantilever. The Grand Hall serves as an interface to the Grande Allée, an urban plaza for the museum's public functions, and a series of gateways into the galleries, courtyard and auditorium. Complementing the quiet reflection of the gallery spaces, a chain of programs along the museum's edge – foyers, lounges, shops, bridges, gardens – offer a hybrid of activities, art and public promenades. Along the way, orchestrated views from a monumental spiral stair and an exterior pop out stair reconnect the visitor with the park, the city, and the rest of the museum. Within the boxes, mezzanines and overlooks link the temporary and permanent exhibition spaces. On top of each of the gallery boxes, roof terraces provide space for outdoor displays and activities. New exhibition spaces are connected to the museum's existing buildings by a 130m (427 ft) long passageway, creating a permanent home for the museum's 40m (132 ft) "Hommage à Rosa Luxemburg" by Jean-Paul Riopelle. Through its sheer length and changes in elevation, the passage creates a surprising mixture of gallery spaces that lead the visitor, as if by chance, to the rest of the museum complex.

The cantilevered structure is supported by a hybrid steel truss system and accommodates galleries uninterrupted by columns. The layered facade is simultaneously structural, thermal and solar, addressing the seemingly contradictory needs of natural light and thermal insulation for Québec's harsh winter climate. The triple layered glass facade is composed of a 2D printed frit that pattern mimics the truss structure, a 3D embossed glass, and a layer of diffuser glass. In the galleries, insulated walls are located behind the translucent glass system, with a gap between that lights the building at night like a lantern in the park. The Grand Hall is enclosed by a glass curtain wall with glass fins that allow virtually unobstructed and inviting views to the Charles Baillairgé pavilion through a glass wall and ceiling. The contrast between the translucent gallery boxes and clear grand hall reinforces the reading of the building's stacking and cantilevering massing.

Espace Sud
Fondation Marcelle et Jean Coutu

Dokk1图书馆

Schmidt Hammer Lassen Architects

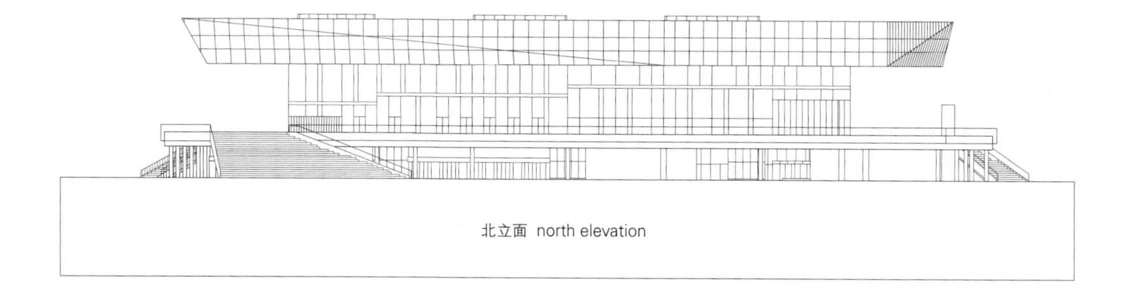

北立面 north elevation

1. 出租办公区
2. 三层开口
3. 多媒体文件收藏区
4. 媒体坡道
5. 会议室
6. 单元自习室
7. 车辆传送间
8. 自动停车系统
9. 艺术品收藏区

1. rentable office area
2. open to level 2
3. media collection
4. media ramp
5. meeting room
6. study cell
7. transfer parking cabin
8. automatic parking system
9. art

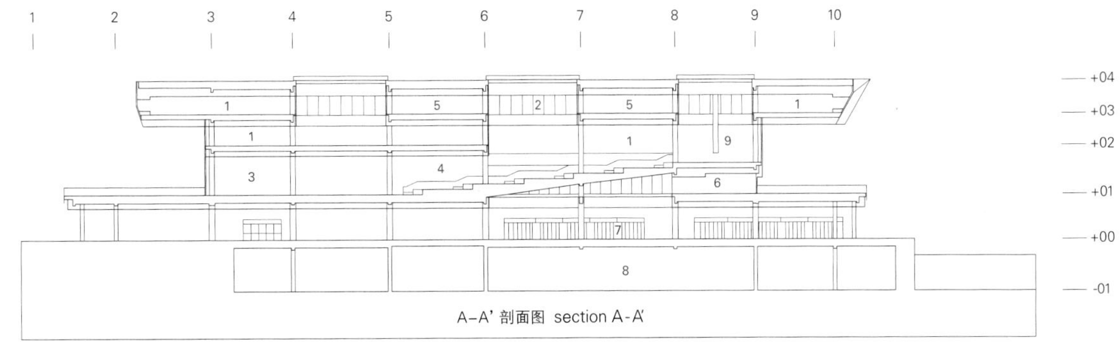

A–A' 剖面图 section A-A'

这座占地 30000m² 的建筑是城市媒体空间项目的一部分，正在改变着奥胡斯海滨。该项目集一座新建图书馆、一个市民服务中心、有一千个停车位的自动停车场和新的临港公共广场于一体。

这座巨大的多边形建筑位于奥胡斯河的河口，将城市与港口和水域连接起来。这一消遣娱乐带的北面和南面都被森林包围，一条新的轻轨线路在此穿过，在 Dokk1 图书馆专门设了一站。道路交通经过重新规划，可以直达该建筑下面的大型地下停车场。在该项目中也设计了新的自行车车道以及超过 450 个自行车停车位。Dokk1 图书馆将成为一个新的枢纽，它可被比喻成"铰链"，开启了奥胡斯市的崭新体验。

Dokk1 图书馆的外形像是一个多边形切割体漂浮在透明的玻璃建筑之上，而下面是带有雕塑般的大型楼梯的裙楼。楼梯成扇形散开，与街面和建筑四周新修的海港人行漫步道连接。该玻璃大楼经过设计，成为一个开放的城市空间，可以 360 度欣赏水景、海港、森林和城市。建筑顶部是一个多边形切割体，给人的印象一直在旋转和移动，因而没有严格意义上的正面或者背面。建筑外立面由多孔金属网板构成。城市和港口的规模都在设计上得到体现，这些金属网板时而扭曲、时而中断、时而转向。立面设计的概念参照了起重机和轮船这种大型机械，力图与周围建筑的高度和港口的规模产生互动。

多边形的顶层包含城市管理办公室以及出租的办公区域。媒体坡道将开放明亮的图书馆两层连接起来。媒体坡道包括五个平台，每一个都有专门用途：展览区、游戏区、互动工作坊、阅读区以及活动区。媒体坡道与文学和媒体区域相结合，提供了一条蜿蜒的活动路线，引导人们一路向上，到达顶层的儿童活动区。

Dokk1

The 30,000 square meter building is part of the urban mediaspace project, which is transforming the harbor front of Aarhus. The project houses the new library, a citizen service center, automatic parking for 1,000 cars and new harbor-side public squares.

The large polygonal building is located at the mouth of the Aarhus River, connecting the city with the port and water. Flanked by forests to the north and south, a recreational belt has been created through which a new light railway will run with a dedicated station at Dokk1. Road traffic will be redirected, so that it has direct access to the large under-ground car park beneath the building. New cycle paths run through the scheme that includes over 450 bicycle parking spaces. Dokk1 will become a new hub and metaphorically the "hinge" that

opens up to new experiences in Aarhus city.

Dokk1 is designed as a polygonal slice that hovers above a glazed building resting on a podium with large sculptural stairs. The stairs fan out to street level and the new harbor promenade is surrounding the building. The glass building is designed as an open urban space with 360-degree views of the water, harbor, forest and city. The building has no clear front or back, which is emphasized by the multi-edged top slice that creates the impression of rotation and movement. The facade is made of expanded metal, the scale of the city and harbor is reflected in the design, which twists, breaks and turns. The concept of the facade design is an interaction with the height of the surrounding buildings and the scale of the port with reference to large elements, such as cranes and ships.

The polygonal top floor contains the municipal administration offices along with rental offices. The two open levels of the library are connected by the media ramp, which consists of five platforms, each dedicated to an activity: exhibitions, gaming, interactive workshops, reading and events. Combined with the literature and media sections, the media ramp offers a meandering path of activity through the building leading up to the children's area on the top floor.

DOKK

Levi's

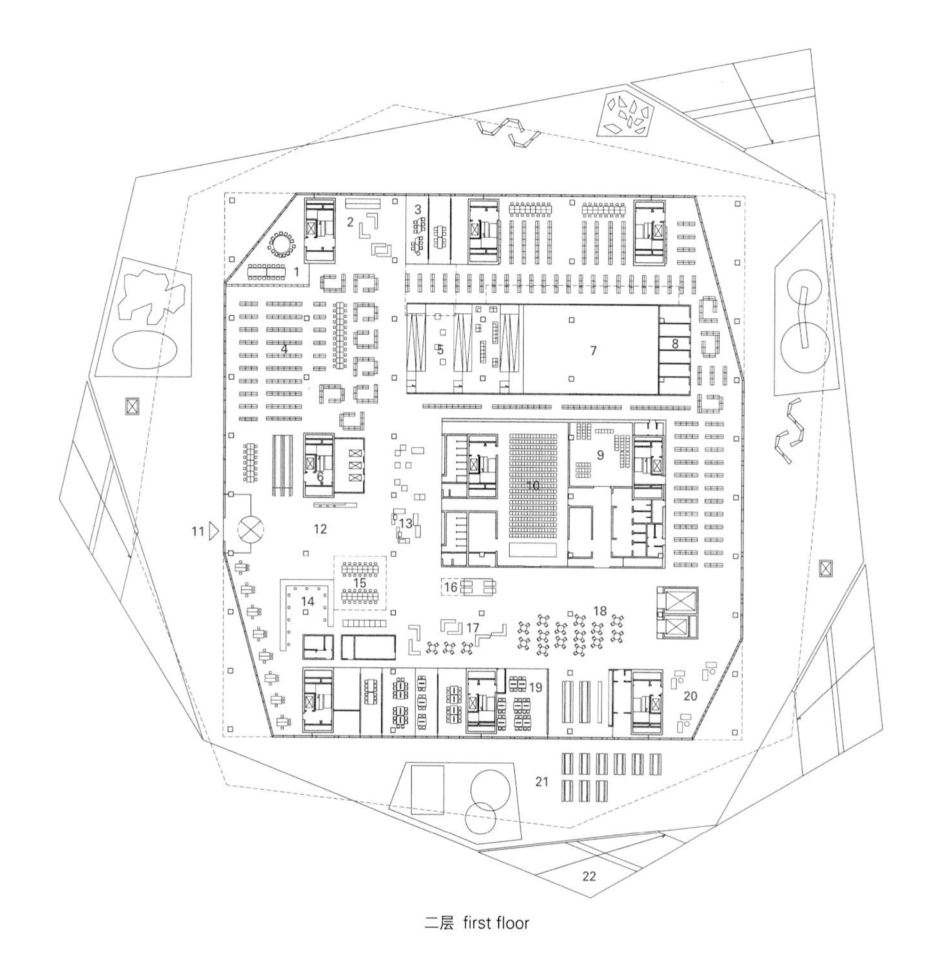

二层 first floor

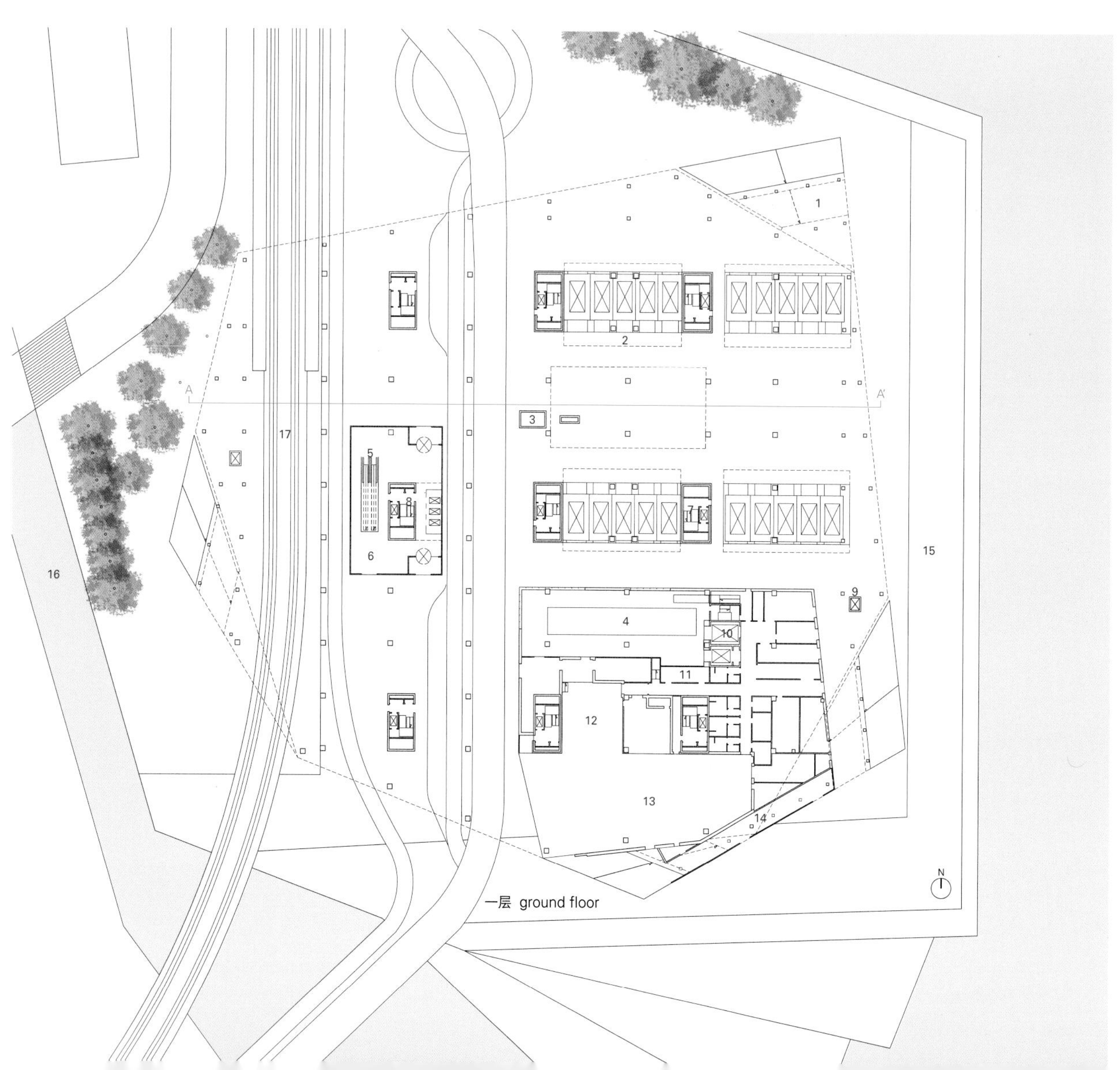

一层 ground floor

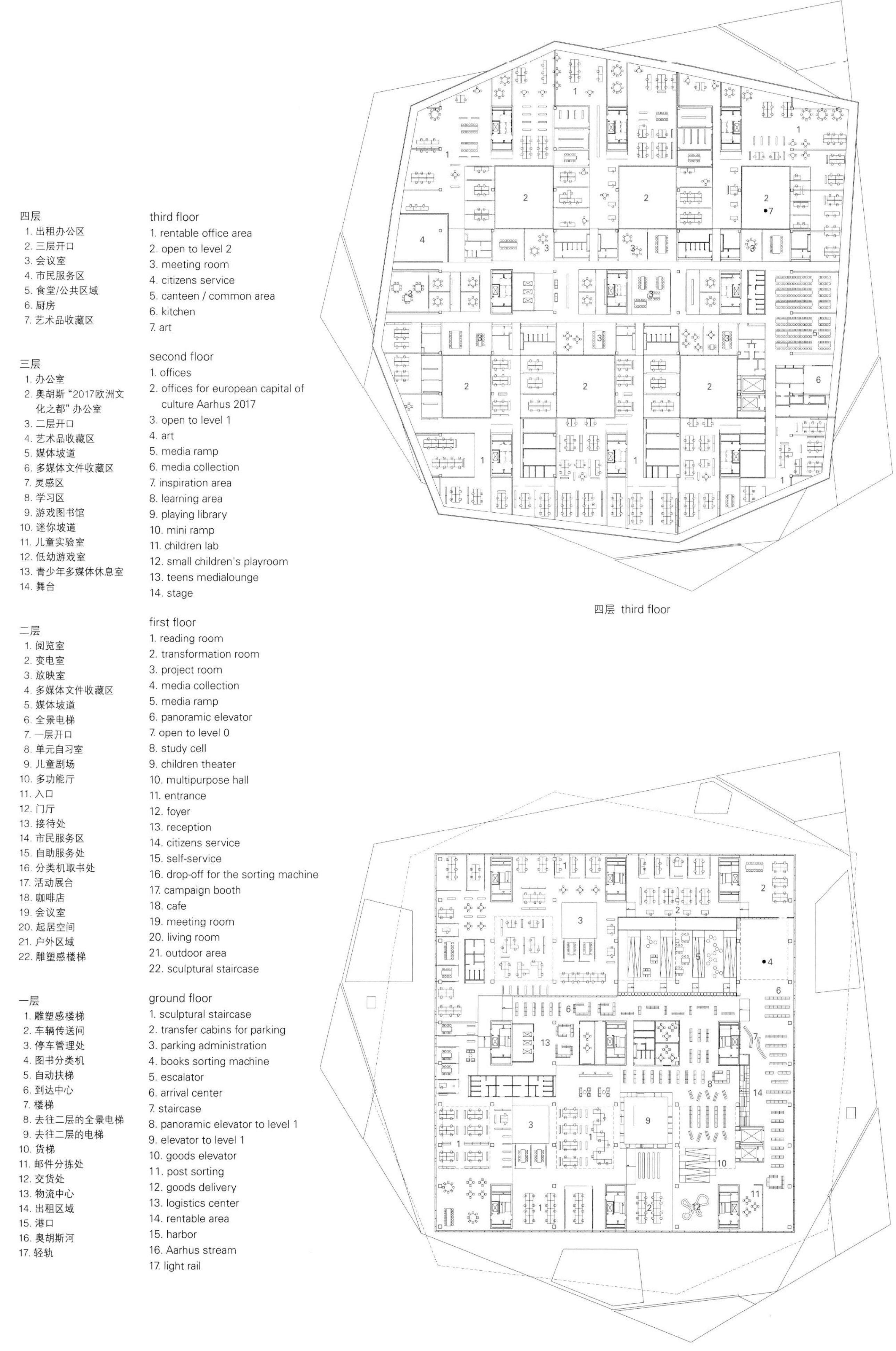

四层

1. 出租办公区
2. 三层开口
3. 会议室
4. 市民服务区
5. 食堂/公共区域
6. 厨房
7. 艺术品收藏区

third floor

1. rentable office area
2. open to level 2
3. meeting room
4. citizens service
5. canteen / common area
6. kitchen
7. art

三层

1. 办公室
2. 奥胡斯"2017欧洲文化之都"办公室
3. 二层开口
4. 艺术品收藏区
5. 媒体坡道
6. 多媒体文件收藏区
7. 灵感区
8. 学习区
9. 游戏图书馆
10. 迷你坡道
11. 儿童实验室
12. 低幼游戏室
13. 青少年多媒体休息室
14. 舞台

second floor

1. offices
2. offices for european capital of culture Aarhus 2017
3. open to level 1
4. art
5. media ramp
6. media collection
7. inspiration area
8. learning area
9. playing library
10. mini ramp
11. children lab
12. small children's playroom
13. teens medialounge
14. stage

二层

1. 阅览室
2. 变电室
3. 放映室
4. 多媒体文件收藏区
5. 媒体坡道
6. 全景电梯
7. 一层开口
8. 单元自习室
9. 儿童剧场
10. 多功能厅
11. 入口
12. 门厅
13. 接待处
14. 市民服务区
15. 自助服务处
16. 分类机取书处
17. 活动展台
18. 咖啡店
19. 会议室
20. 起居空间
21. 户外区域
22. 雕塑感楼梯

first floor

1. reading room
2. transformation room
3. project room
4. media collection
5. media ramp
6. panoramic elevator
7. open to level 0
8. study cell
9. children theater
10. multipurpose hall
11. entrance
12. foyer
13. reception
14. citizens service
15. self-service
16. drop-off for the sorting machine
17. campaign booth
18. cafe
19. meeting room
20. living room
21. outdoor area
22. sculptural staircase

一层

1. 雕塑感楼梯
2. 车辆传送间
3. 停车管理处
4. 图书分类机
5. 自动扶梯
6. 到达中心
7. 楼梯
8. 去往二层的全景电梯
9. 去往二层的电梯
10. 货梯
11. 邮件分拣处
12. 交货处
13. 物流中心
14. 出租区域
15. 港口
16. 奥胡斯河
17. 轻轨

ground floor

1. sculptural staircase
2. transfer cabins for parking
3. parking administration
4. books sorting machine
5. escalator
6. arrival center
7. staircase
8. panoramic elevator to level 1
9. elevator to level 1
10. goods elevator
11. post sorting
12. goods delivery
13. logistics center
14. rentable area
15. harbor
16. Aarhus stream
17. light rail

四层 third floor

三层 second floor

TOILETTER
TOILETTER

ÆSESAL

项目名称：Dokk1
地点：Aarhus, Denmark
建筑师：schmidt hammer lassen architects
总建筑师：schmidt hammer lassen architects
工程师：Alectia Rådgivende Ingeniører
景观设计：Arkitekt Kristine Jensens Tegnestue
客户：Aarhus Municipality
造价：EUR 280 million excluding VAT
面积：35,600m²
设计时间：2009 / 竣工时间：2015
摄影师：©Adam Mørk (courtesy of the architect)

葡萄酒之城博物馆

XTU Architects

我们如何才能使设计与该地区的精神保持一致？这儿以前的泊船坞与这个城市的工业遗产密切相关。

该地属于河流区域。水是它的生命力，大自然的力量重新塑造着这些人工河堤。当宽阔的河流缓慢蜿蜒地流向远方时，其中的一个河湾就能形成一个特别的要塞：成为港口的入口，即这座城市的门户。这样的地方需要一个有亲切感的标志性建筑物来守护，也就是这个港口和这座城市的守护天使。她是楔石，是水平天际线上的直立的图标，是一座从四面八方都能看到的灯塔 ，居高临下俯瞰着河的两岸。

建筑的形状像古老的卵石一样光滑、圆润，轻盈的线条庄严地向水边扭转，体现了流动性的精髓。

建筑外表装饰着灯带，闪闪发光。随着视线的移动，灯光的颜色可由紫色变成金色。建筑物看起来虚无飘渺、神秘莫测，从河流的迷雾中庄严升起。设计受到了该地区精神的启发，向该地区的葡萄酒精神致敬。

登上宽阔的楼梯，参观者就可到达论坛大厅。宽阔的楼梯缓缓地蜿蜒向上通到展览厅。但在楼梯口，参观者就可以直接看到展览厅，并且很快就能意识到，这样的设计布局是为了让人们想到“漩涡”的形状——就像人们在品尝葡萄酒之前会转一转酒杯来醒酒。游客看到的一切都在表现这种运动——形式、光、图形和文字。因此，展览区是按照环形路线来布置的：游客从论坛大厅开始，在游览完没有任何棱角的流动空间之后，最后再回到原处。加伦河静静地在该建筑的四周流淌，透过玻璃幕墙，能够欣赏到她壮丽的景色。建筑内部再现了这种流动感，但原来的怪石险滩似乎早已经被流水磨平、磨圆。

游客总能意识到波尔多葡萄酒的存在。透过窗户，一览无余的视野提醒着游客他们身处何处：这是一座由葡萄酒主宰一切的城市，也是开启探索葡萄酒世界之旅的出发点。

Cité du Vin

How can we fit in with the spirit of the area? The former wet docks are so intimately associated with the city’s industrial heritage.

This place belongs to the river. The water is its life force, the power of nature reclaiming these man-made banks. As the broad river winds gently towards its destiny, one of its curves forms a special stronghold: the entrance to the port, the gateway to the city. Such a place needs to be watched over by a benevolent landmark, a guardian angel for the port and the city. A keystone, a vertical icon on this horizontal skyline, a beacon that can be seen far and wide, commanding the banks of the river.

A shape as smooth and rounded as an ancient pebble, weightless lines rolling majestically down towards the water.

BASSIN À FLOT N°2
BASSIN À FLOT N°1
Fleuve Garonne

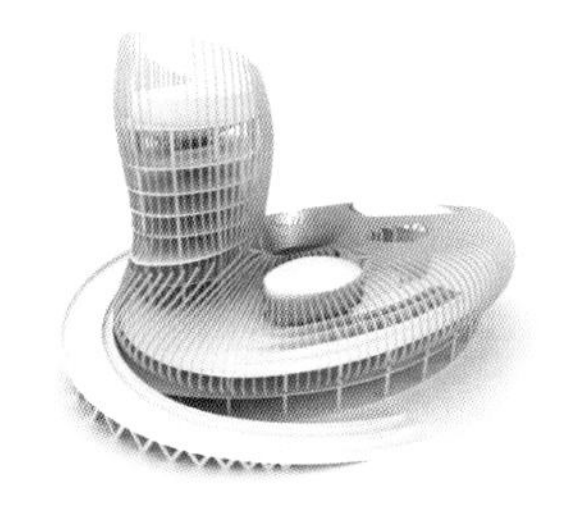

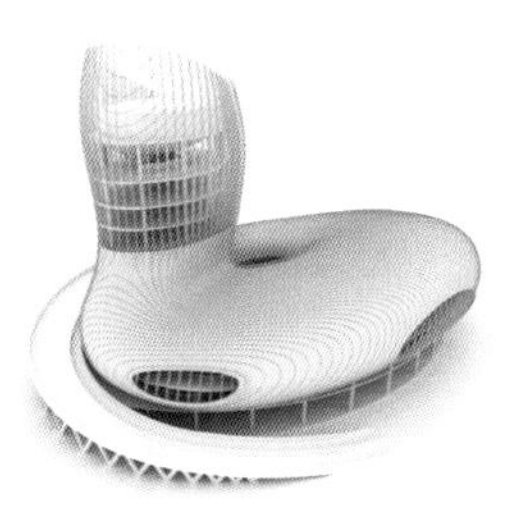

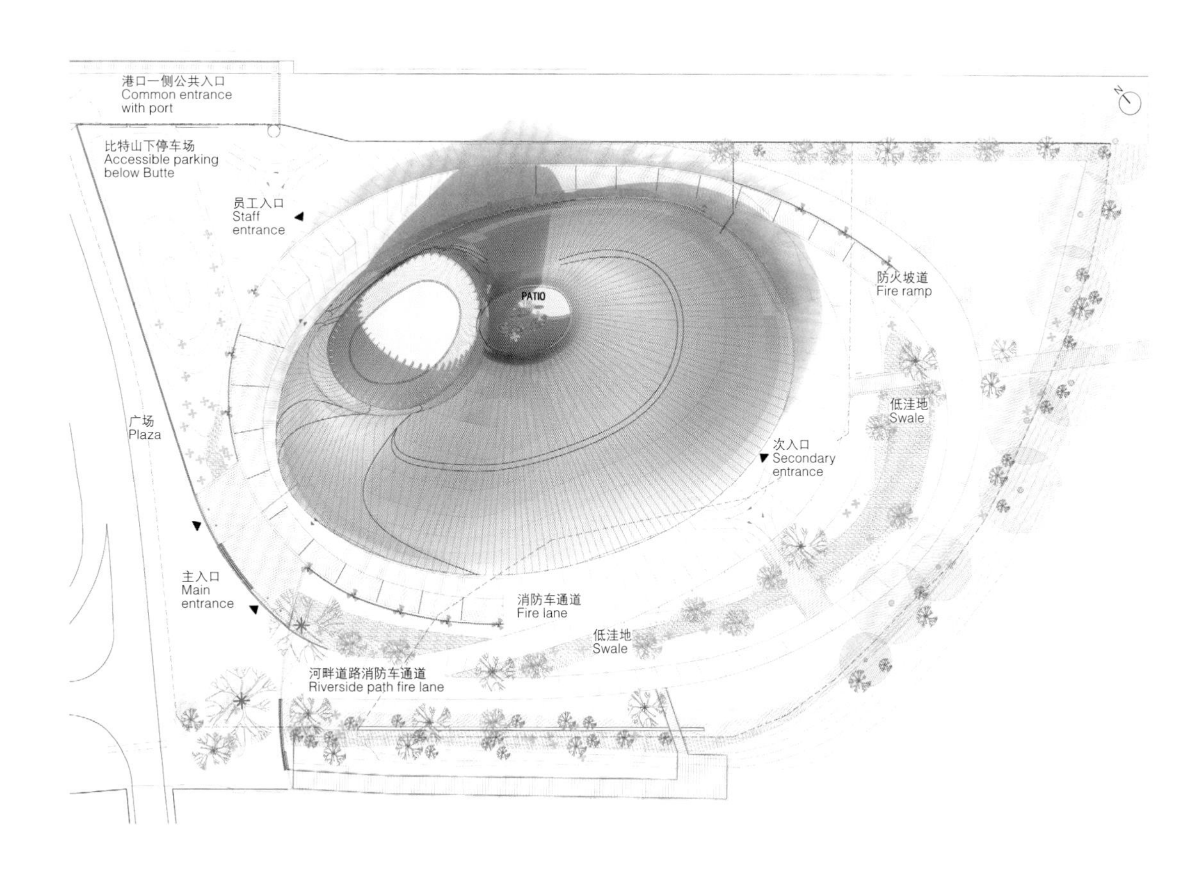
港口一侧公共入口
Common entrance with port
比特山下停车场
Accessible parking below Butte
员工入口
Staff entrance
PATIO
防火坡道
Fire ramp
低洼地
Swale
广场
Plaza
次入口
Secondary entrance
主入口
Main entrance
消防车通道
Fire lane
低洼地
Swale
河畔道路消防车通道
Riverside path fire lane

La Cité des civilisatio
du vin à Bordeaux
30
10
LM

The essence of fluidity.
This form is clothed in swathes of light, shimmering and fluctuating from purple to gold as the viewer moves. Ethereal, enigmatic, atmospheric, rising magnificently from the mists of the river. Inspired by the spirit of the place, dedicated to the spirit of wine.
The visitors arrive at the Forum by climbing the sweeping staircase, curving gently upwards to the exhibition. At the head of the stairs, visitors will be faced immediately with the exhibition, and will soon realize that the layout is designed to evoke the idea of a "whirlwind" - like the circular movement which awakens a wine before tasting. Everything the visitors see will reflect this motion - the forms, the light, the images and texts. The exhibition is thus laid out along a circular route: visitors start at the Forum and eventually arrive back there, after a journey through a flowing space with no corners.
The glass walls offer a spectacular view of the river Garonne, flowing serenely around the building. Inside we have recreated this sense of flow, running serenely between obstacles which seem to have been smoothed and rounded by the passing water.
Visitors are always conscious of the presence of Bordeaux. Through the windows, the panoramic view reminds visitors where they are: in a city where wine reigns supreme, and which provides the point of departure for a voyage into the world of wine.

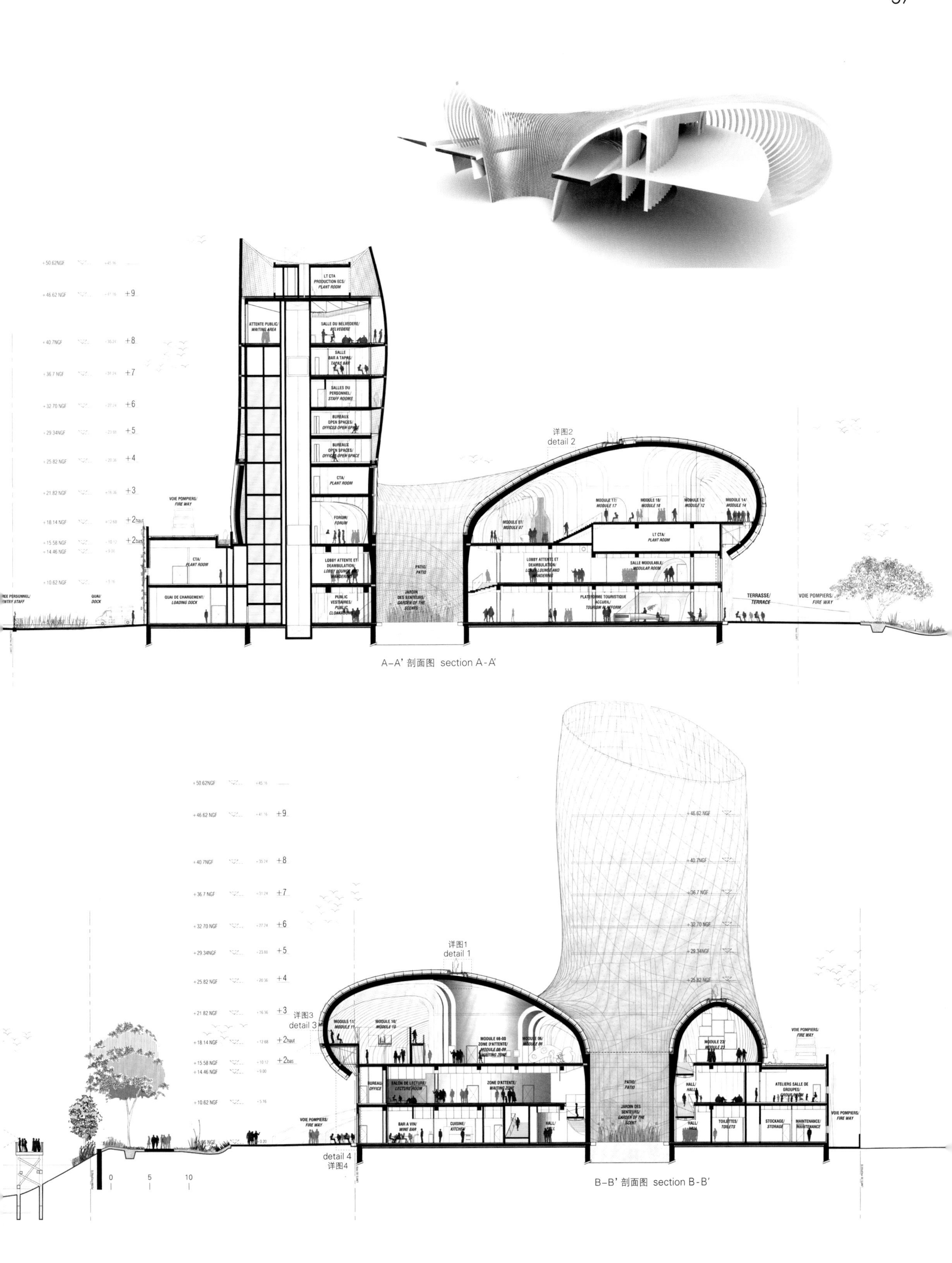

A-A' 剖面图 section A-A'

B-B' 剖面图 section B-B'

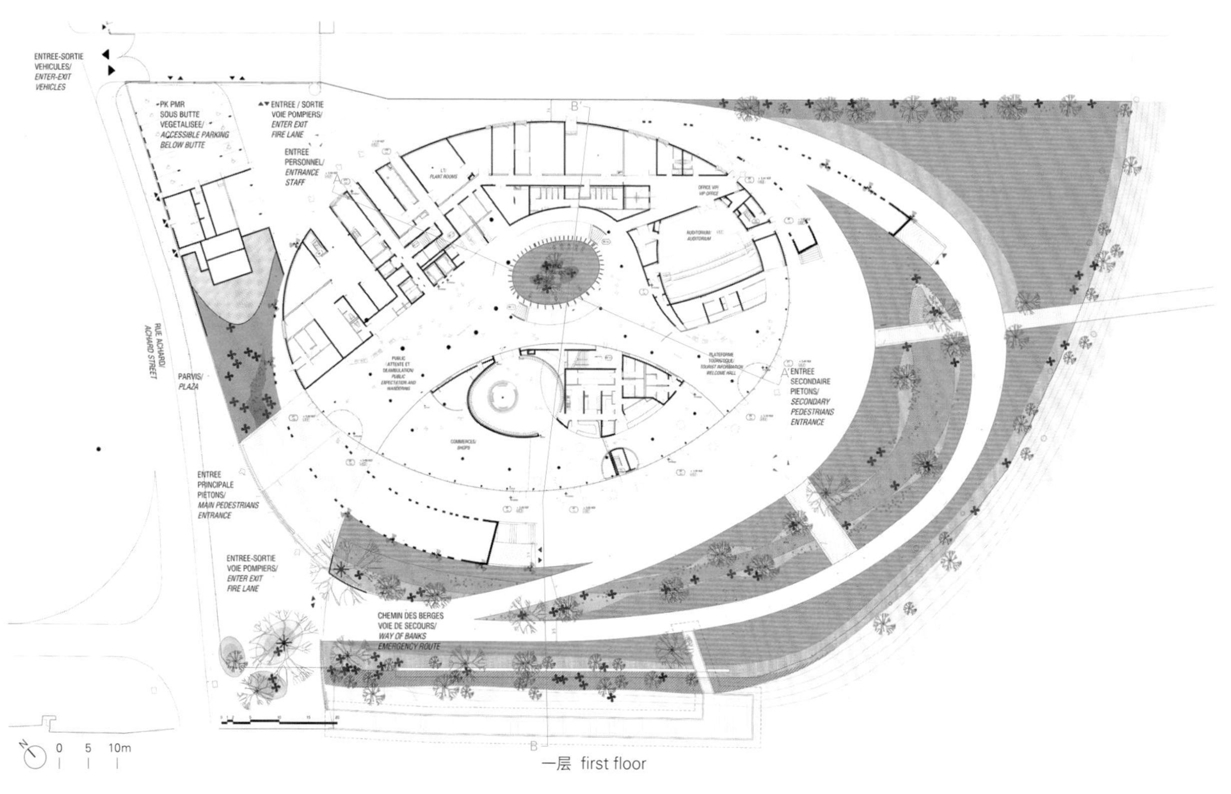

一层 first floor

项目名称：Cité du Vin / 地点：134 - 150 Quai de Bacalan, 33300 Bordeaux, France / 建筑师：Anouk Legendre + Nicolas Desmazières_XTU architects / 项目负责人：Mathias Lukacs, Dominique Zentelin / 现场团队：Delphine Isart, Claire Leroux, Thibault Le Poncin, Joan Tarragon / 研究团队：Joan Tarragon, Gaëlle Le Borgne, Stefania Maccagnan, Cristina Sanchez / 布景设计：Casson Mann / 结构与流体工程：SNC-Lavalin / 环境工程：Le Sommer Environnement / 景观设计师：Sequences Paysage (Camille Jullien) / 客户：Ville de Bordeaux / 用地面积：13,644m² / 建筑面积：12,927m² (usable surface included 2,800m² of permanent exhibition) / 竣工时间：2016.5 / 摄影师：courtesy of the architect-p.56, p.58, p.59, p.64~65; ©ANAKA (courtesy of the architect)-p.54~55, p.62~63; ©Delphine Isart (courtesy of the architect)-p.52; ©Julien Lano (courtesy of the architect)-p.50~57, p.60; ©Patrick Tourneboeuf (courtesy of the architect)-p.61 bottom

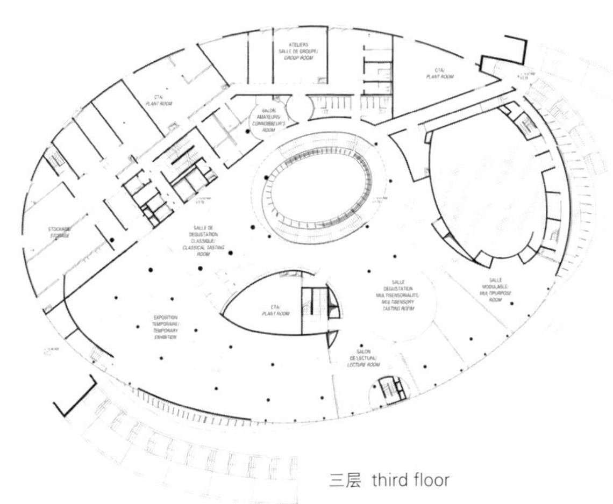

三层 third floor

八层 eighth floor

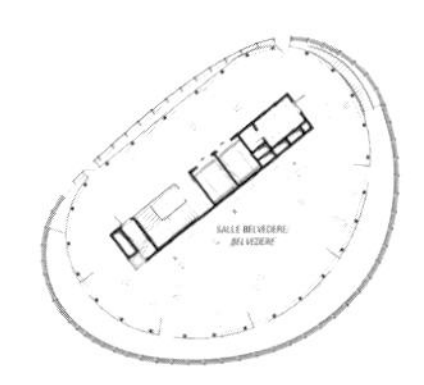

十层 tenth floor

六层 sixth floor

九层 ninth floor

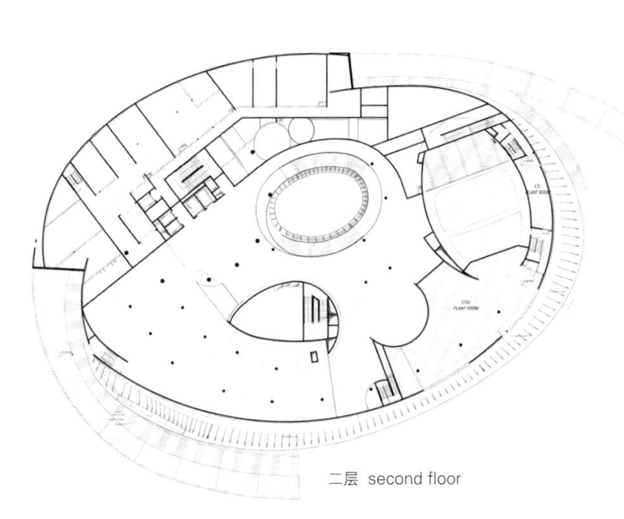

二层 second floor

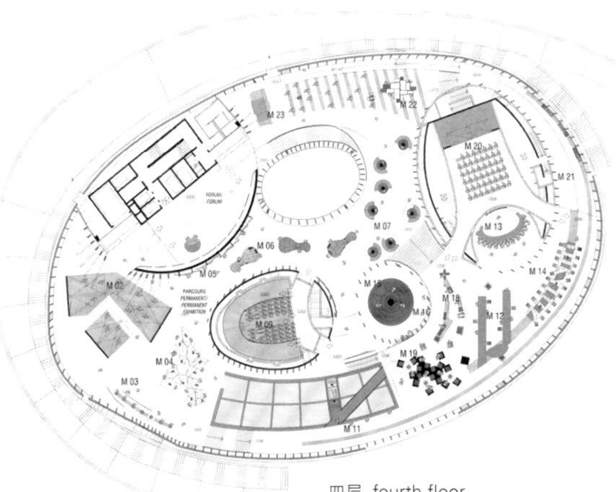

四层 fourth floor

Le cuvier

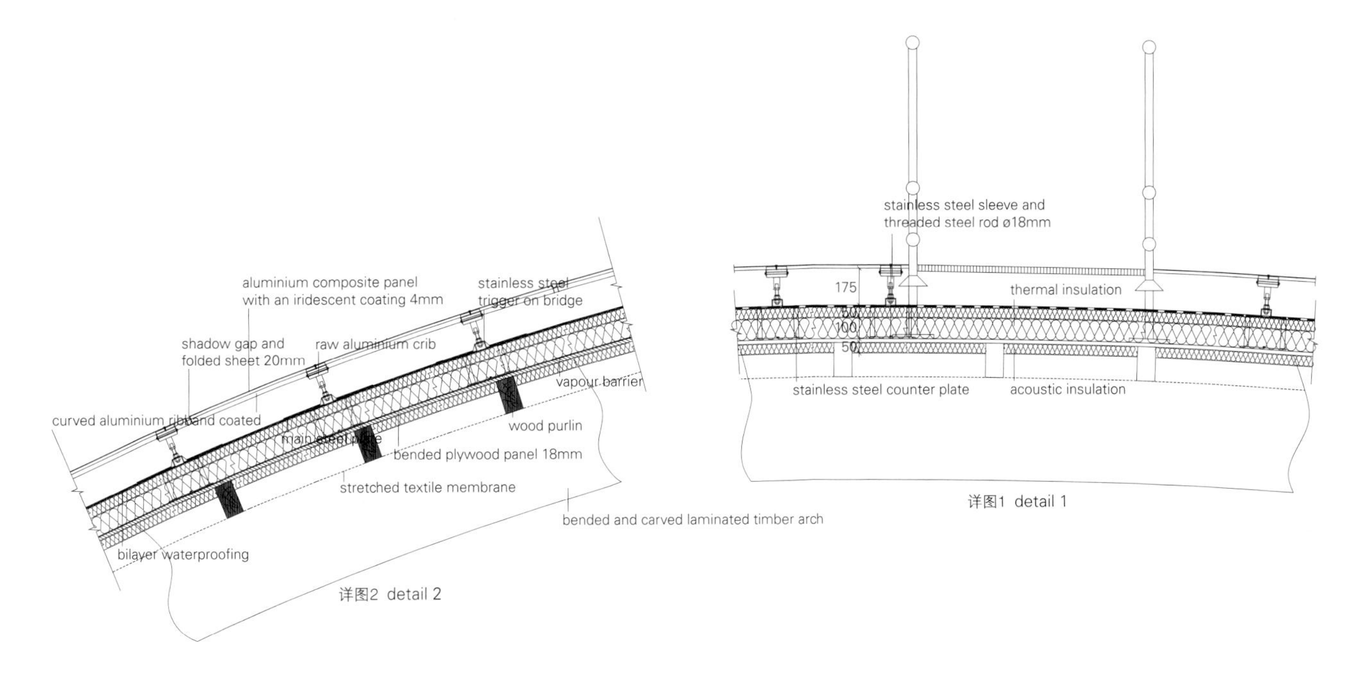

详图1 detail 1

详图2 detail 2

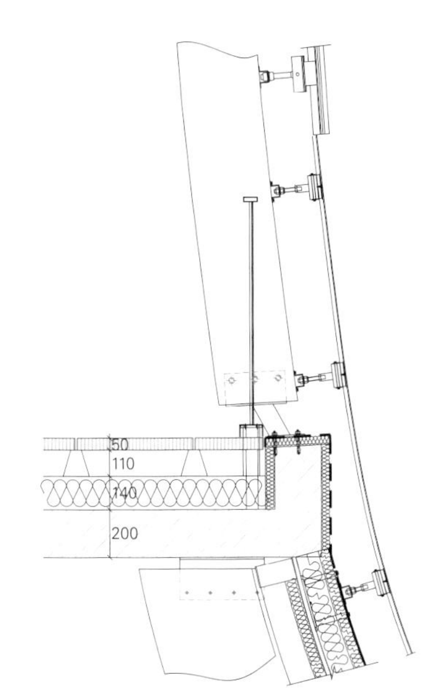

详图3 detail 3

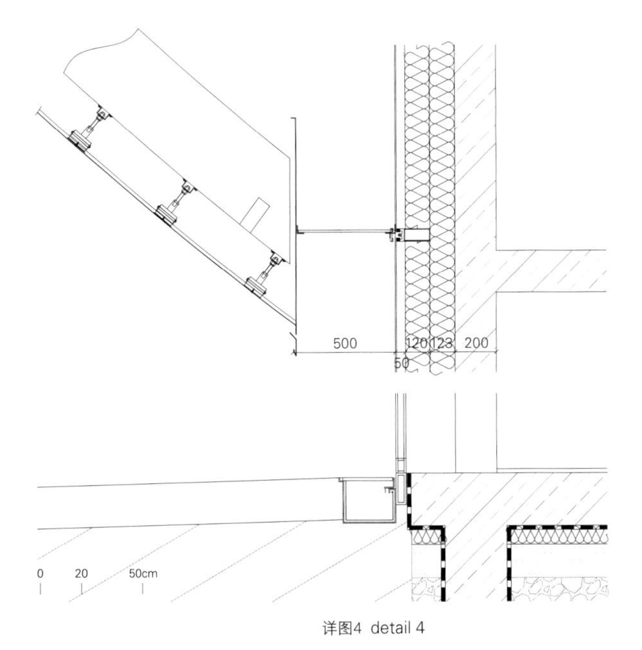

详图4 detail 4

斯塔夫罗斯·尼亚尔霍斯基金会文化中心

Renzo Piano Building Workshop

斯塔夫罗斯·尼亚尔霍斯基金会文化中心位于雅典南部海岸城市卡利地亚，距离雅典城南部 4km。建筑场地位于一座 17 万平方米的景观园区内，包含希腊国家图书馆和希腊国家大剧院。这里曾经是赛马场，现在是 2004 年奥运会留下的一个停车场。一旦项目建成，将重新找回这里与城市和海洋的联系。

卡利地亚作为希腊法力罗湾最早的海港之一，一直以来都与水有着牢不可破的联系。而现在，虽然场地邻近大海，但是在这里却看不到任何海洋的景色。为了找回与海洋的联系，在场地南侧（朝海一侧）修建了一个人工山坡。这一坡地公园的最高处就是文化中心建筑，在楼内可欣赏到壮观的海上风光。

歌剧院与图书馆被合并在一栋建筑中，用"城市广场"（Agora）这样的希腊传统公共空间将这两个主要设施连接在一起。歌剧院由两个礼堂组成，其中一个 450 座的礼堂用于传统歌剧和芭蕾演出，另一个 1400 座的礼堂则用于更具有实验性的演出。图书馆不仅是学习和传承文化的场所，也是一种公共资源，一个真正可以实现文化共享和赏析的场所。

全玻璃落地窗的图书馆阅览室位于建筑最上层，上方覆盖着一个巨大的悬挑屋顶。在这个犹如俯卧在地面的透明方盒子里，可 360 度欣赏到雅典市容与大海的壮观美景。场地在视觉上和实际上与水的联系通过公园中一条新修的运河继续延伸，运河旁是一条南北向主要人行道轴线，被称为海滨大道。华盖般的屋顶既满足了必要的遮阳需要，上面还安装了 1 万平方米的光伏电池，可以产生 1.5 兆瓦的电能供图书馆与歌剧院使用。这一巨大的光伏电池屋顶可以保证建筑在正常开放时间能源方面的自给自足。另外，在所有地方也尽可能地利用了自然通风措施。

与水的视觉联系延伸到公园中，聚焦在海滨大道边的运河上。海滨大道是一条南北向的主要人行道轴线。该建筑综合体的设计目的是获得 LEED 白金级认证。

Stavros Niarchos Foundation Cultural Center

The Stavros Niarchos Cultural Center is located in Kallithea, 4 km south of central Athens. The site comprises the National Library of Greece and the Greek National Opera in a 170,000 sqm landscaped park. The site was once a racetrack and currently a parking lot left over from the 2004 Olympic Games. The project restores the site's lost connections with the city and the sea.

As one of Athens' earliest seaports on Faliro Bay, Kallithea has always had a strong relationship with the water. At present, however, despite its proximity, there is no view of the sea from the site. To restore this, an artificial hill is created at the south (seaward) end of the site. The sloping park culminates

©Ruby on Thursdays (courtesy of the architect)

东南立面 south-east elevation

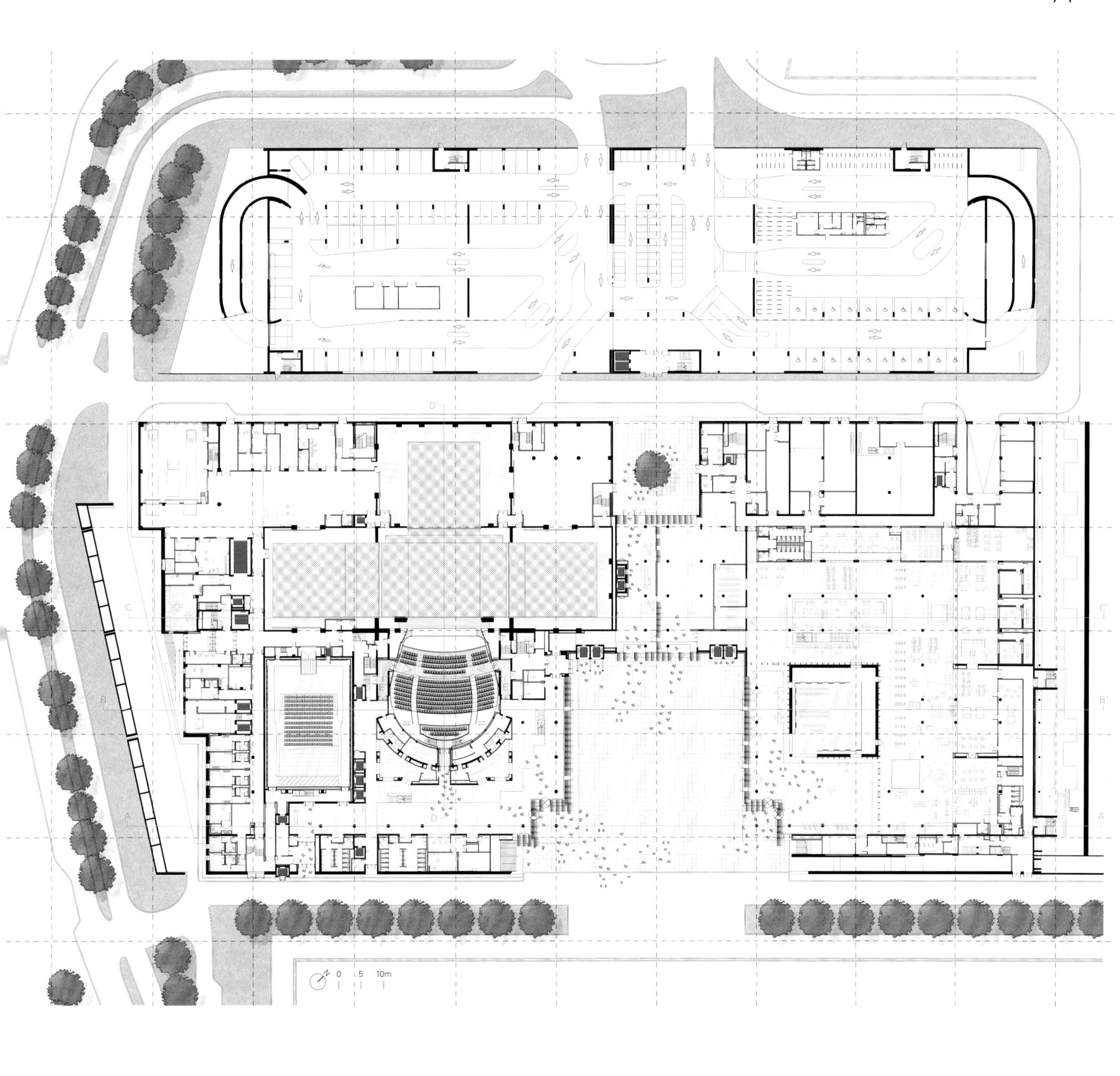

0
5
10m

in the cultural center building, giving it spectacular views towards the sea.

Both opera and library are combined in one building, with a public space, known as the Agora, providing access and connections between the two main facilities. The opera wing is composed of two auditoria, one (450 seats) dedicated to traditional operas and ballets, the other (1,400 seats) for more experimental performances. The library is intended as not only a place for learning and preserving culture, but also as a public resource, a space where culture is truly accessible to share and enjoy.

The entirely glass-walled library reading room sits on top of the building just underneath the canopy roof. A square horizontal transparent box, it enjoys 360-degree views of

Athens and the sea. The site's visual and physical connection with water continues in the park with a new canal that runs along a north–south, main pedestrian axis, the Esplanade. The canopy roof provides essential shade and has been topped with 10,000sqm of photovoltaic cells, enough to generate 1.5 megawatt of power for the library and opera house. This field of cells should allow the building to be self-sufficient in energy terms during normal opening hours. Wherever possible, natural ventilation has been used.

The visual connection with the water continues to the park, where it focus on a channel to the side of the Esplanade, the main pedestrian axis of the site, in the north-south direction. The complex is aiming for an LEED platinum rating.

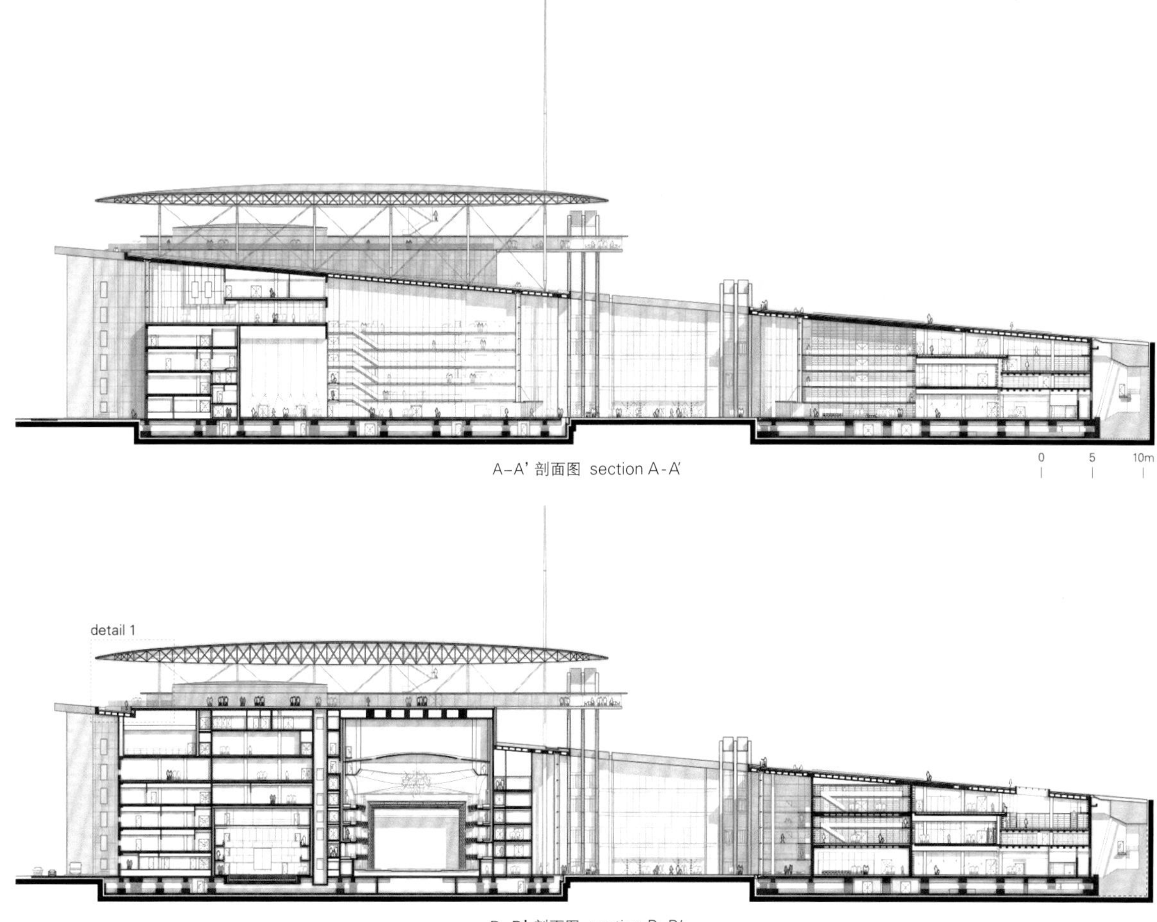

A–A’ 剖面图 section A-A’

B–B’ 剖面图 section B-B’

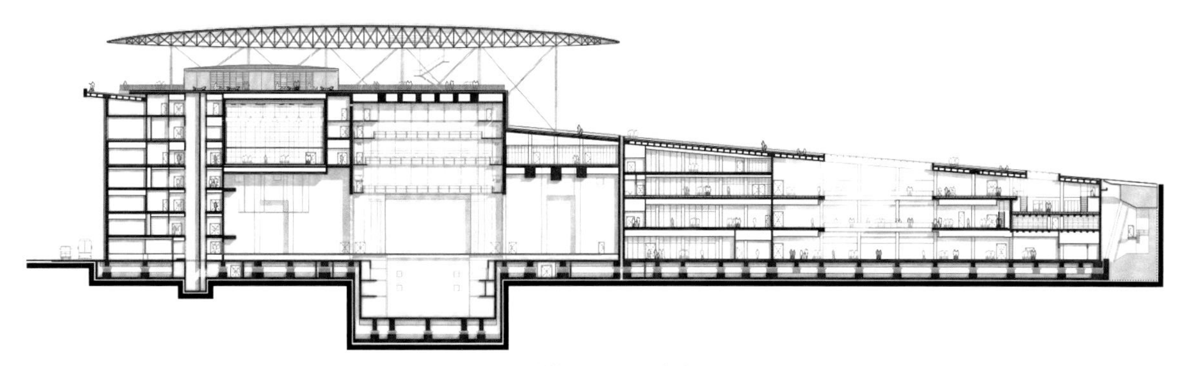

C–C’ 剖面图 section C-C’

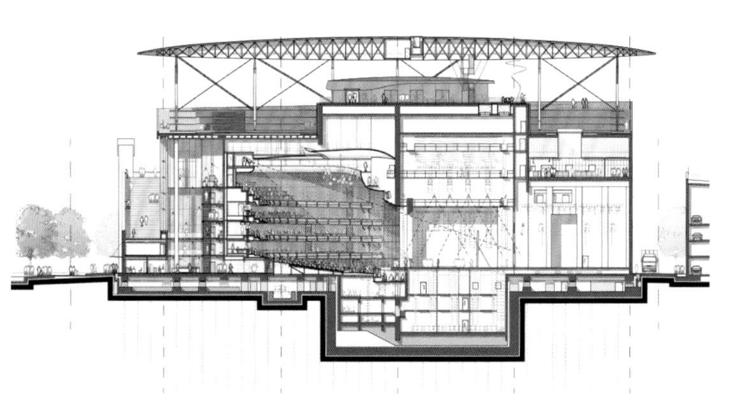

D–D’ 剖面图 section D-D’

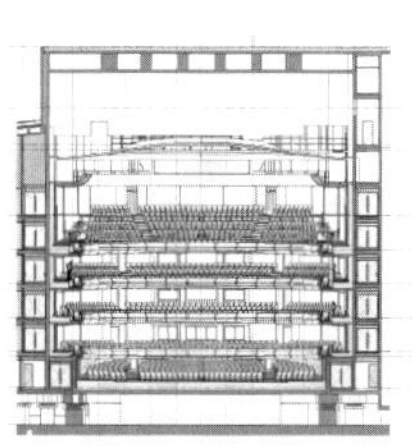

E–E’ 剖面图 section E-E’

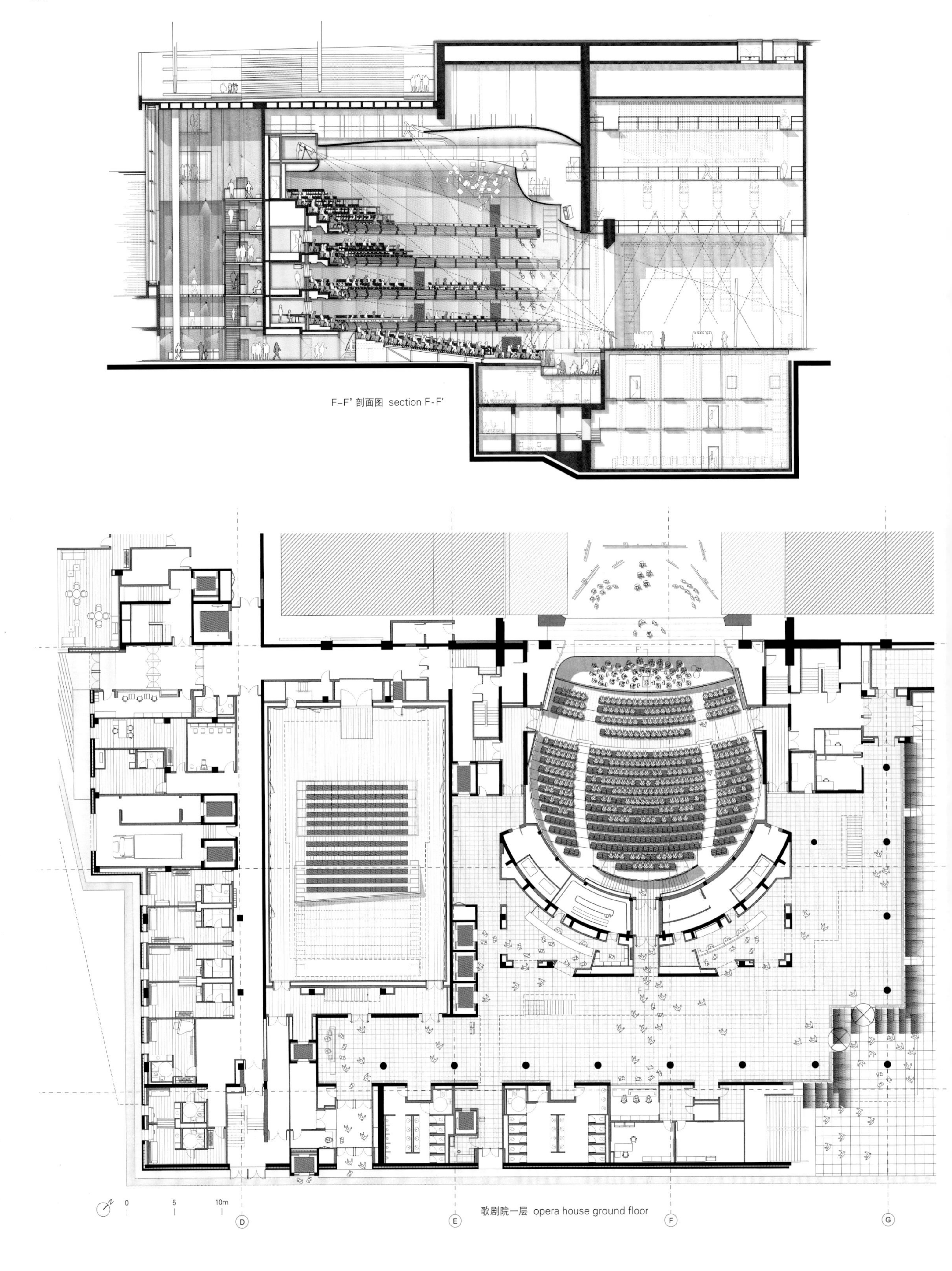

F–F' 剖面图 section F-F'

歌剧院一层 opera house ground floor

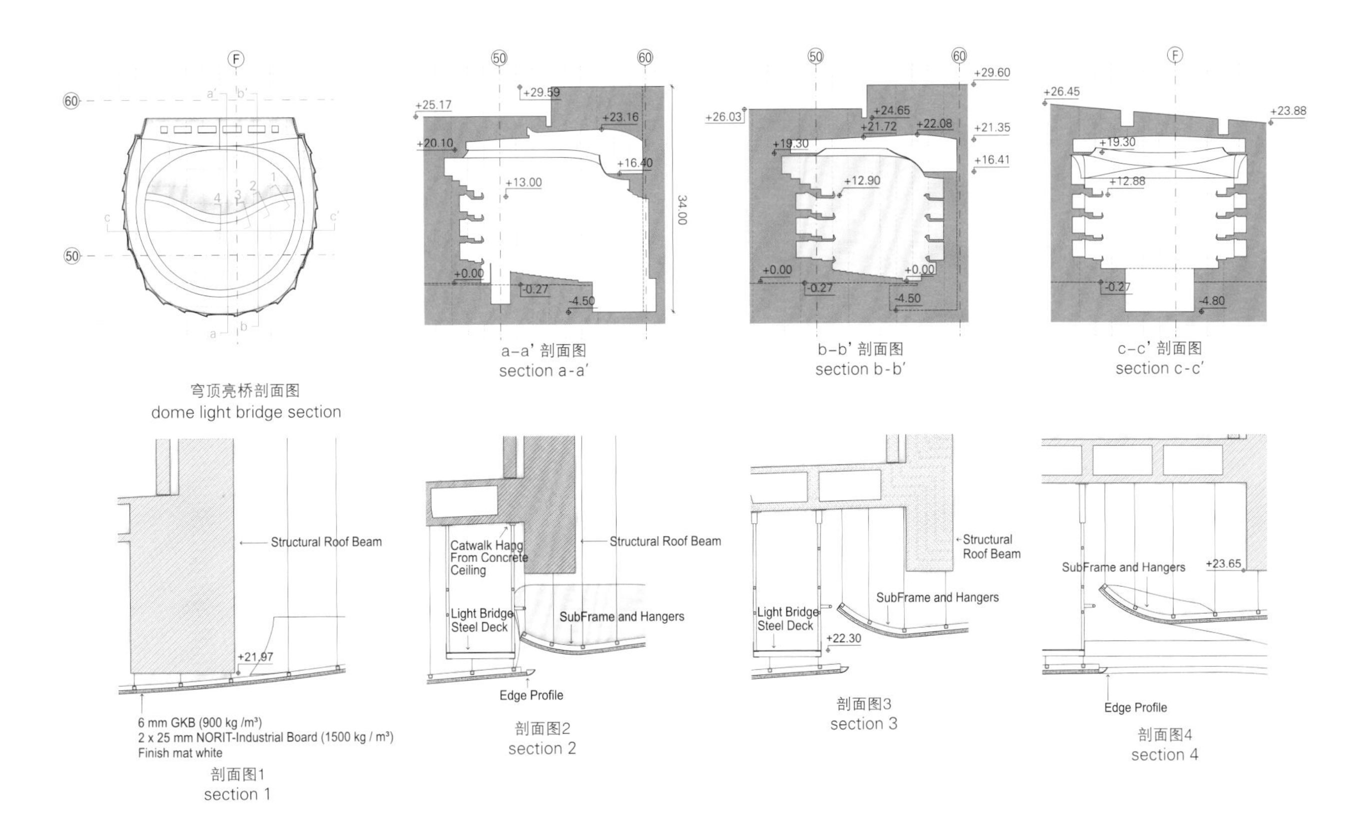

©George Dimitrakopoulous (courtesy of the architect)

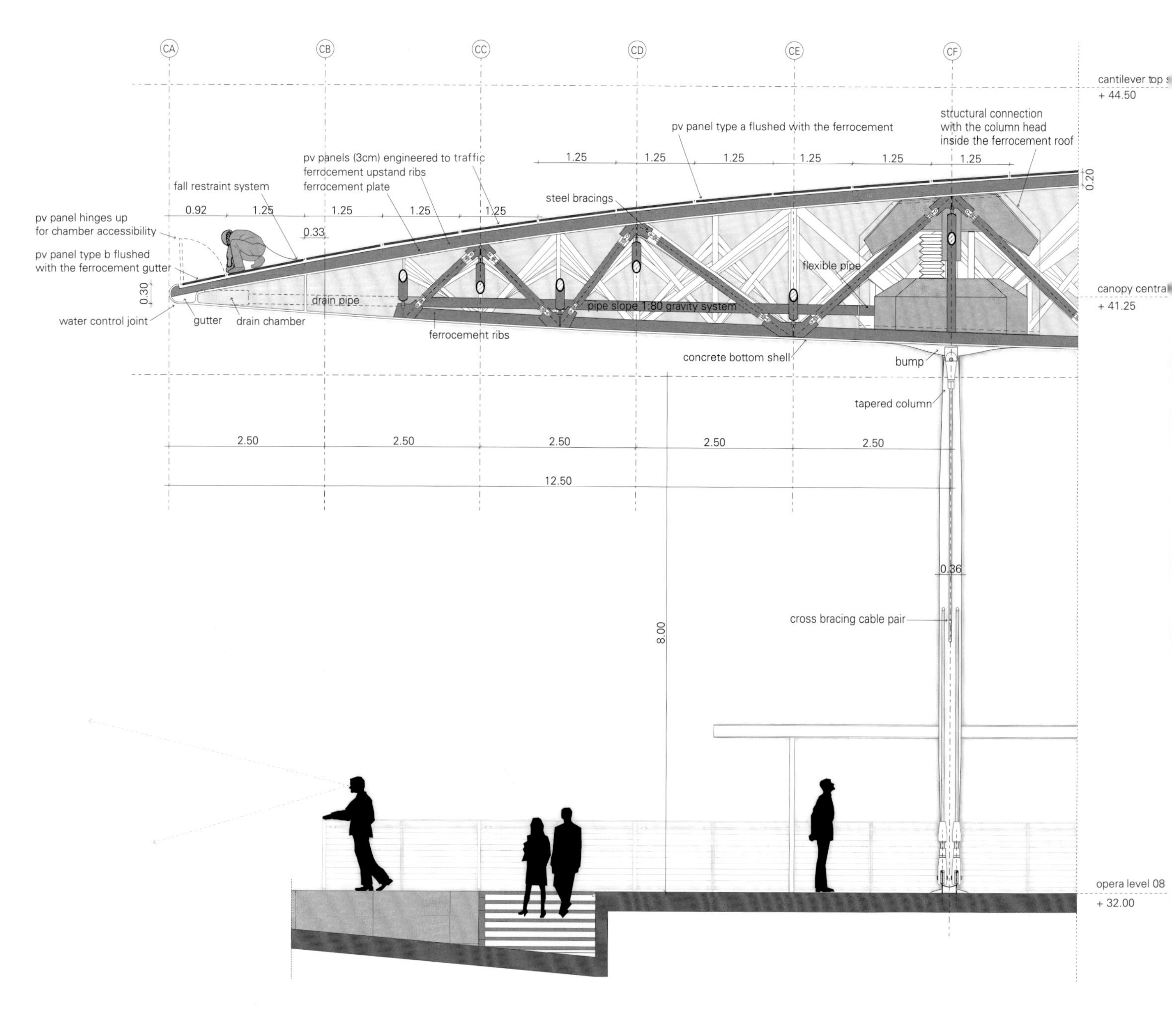

项目名称：Stavros Niarchos Foundation Cultural Center / 地点：Athens, Greece
建筑师：Renzo Piano Building Workshop
项目团队：G.Bianchi, V.Laffineur (partner and associate in charge), S.Doerflinger, H.Houplain, A.Gallissian with A.Bercier, A.Boldrini, K.Doerr, S.Drouin, G.Dubreux, S.Giorgio-Marrano, C.Grispello, M.A.Maillard, E.Ntourlias, S.Pauletto, L.Piazza, M.Pimmel, L.Puech and B.Brady, C.Cavo, A. Kellyie, C.Menas Porras, C.Owens, R.Richardson; S.Moreau; O.Aubert, C.Colson and Y.Kyrkos (models)
合作方：Betaplan (Athens) / 顾问：Expedition Engineering/OMETE _ structure, Arup/LDK Consultants _ MEP, sustainability, acoustics, lighting, security, IT, Theater Project Consultants _ theater equipment, Front _ facade engineering, Deborah Nevins & Associates/H.Pangalou _ landscaping, C&G Partners, M.Harlé/J.Cottencin _ signage, Faithful+Gould _ project and cost management
客户：The Stavros Niarchos Foundation

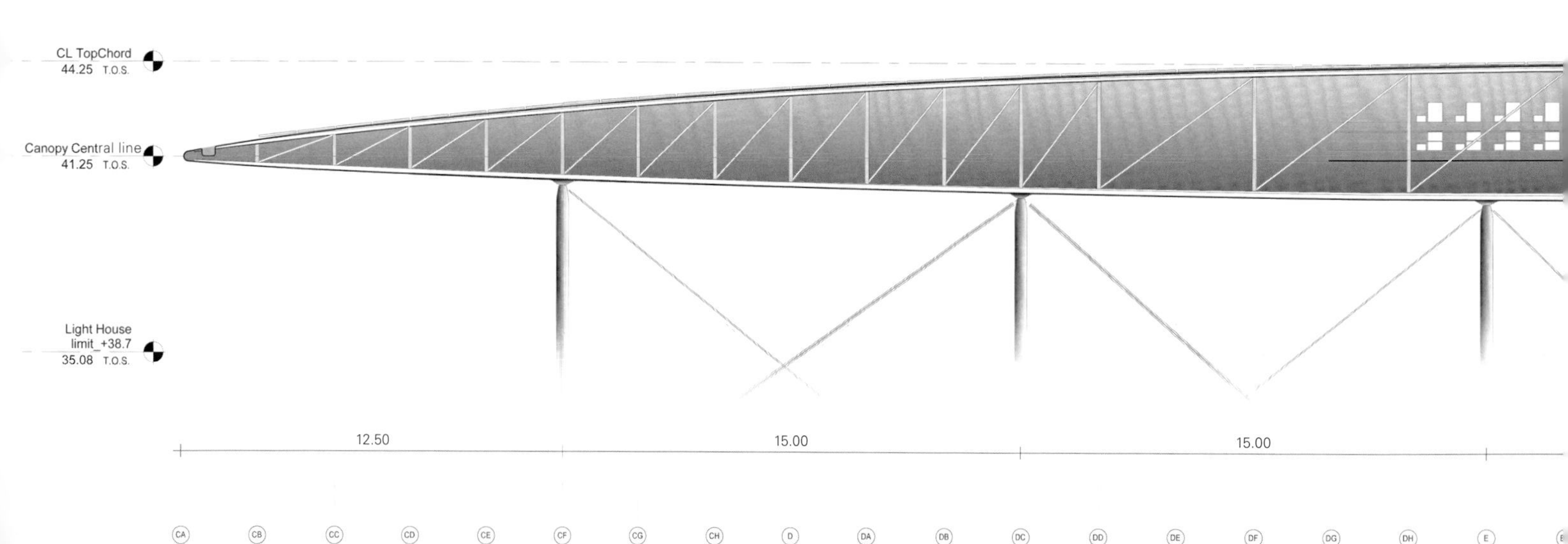

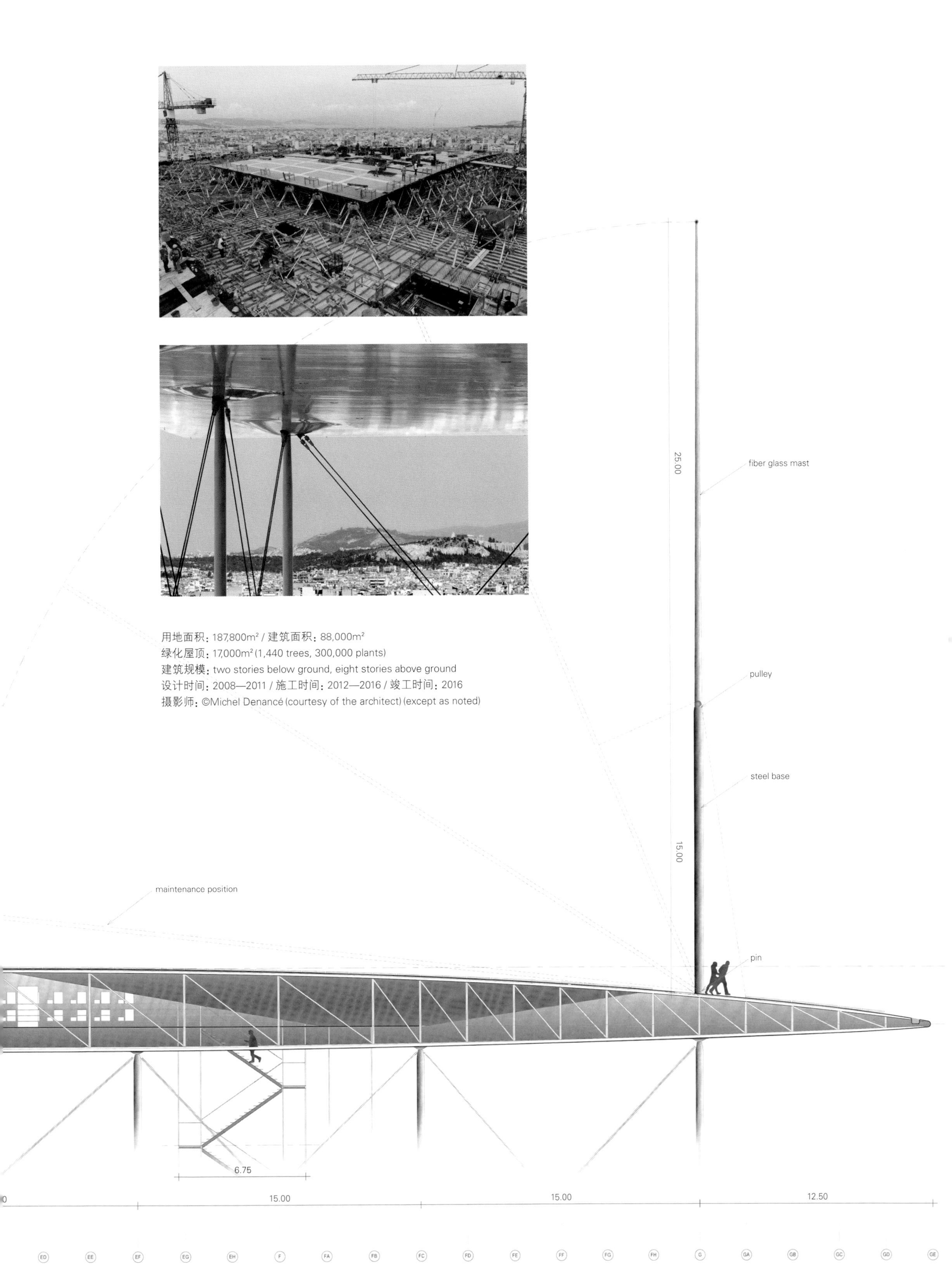

用地面积：187,800m² / 建筑面积：88,000m²
绿化屋顶：17,000m²(1,440 trees, 300,000 plants)
建筑规模：two stories below ground, eight stories above ground
设计时间：2008—2011 / 施工时间：2012—2016 / 竣工时间：2016
摄影师：©Michel Denancé (courtesy of the architect) (except as noted)

详谈社区建筑设计

Narratives for Community Architecture

尽管公共空间的价值如今处于危机时期，但是共享空间和公共空间并没有缩减，反而在扩大。事实上，如今我们拥有的可支配公共空间比以往任何时期都广阔。然而，随着社会冲突的不断上升、世界范围内移民的影响以及全球经济和金融利益的变化，传统的公共空间所承受的压力越来越大，正在致力于平衡其在安全和监督、参与和商业化之间的作用。更糟糕的是，数字化空间的无所不在已经取代了公共空间原先在塑造文化、政治和一般的公民行为方面的主导地位。

本文仔细研究了五个融合了共享、公众和公共空间的综合体案例，这是一些脱离传统公共空间领域而开发的案例。本文探讨了这些项目如何在"公共"空间的定义上提供新的视角（或重新下定义）；以及这些空间的设计者是怎样对社区建筑的多种阐述表示认可的，并且证明了许多中型和小型的社区相结合同样可以为实际的公共环境增加价值。

Despite the current state of crisis concerning the value of public space, shared and common space is not shrinking, but expanding. In fact, we have a vaster volume of communal space to our disposal than ever before. However, with rising social conflict, the effects of worldwide migrations and shifts in global economic and financial interest, the traditional public space is increasingly under pressure and trying to balance its act between security and surveillance, participation and commercialization. To make matters worse, the omnipresence of digital space has overtaken the public space's original position as the prime location for shaping culture, politics and civic conduct in general.

The article takes a closer look at five samples of new hybrids of shared, public and communal space that are developing outside the realm of the traditional public space. It explores how these projects each offer new insights about the definition – and possibly the redefinition - of what makes a place "public"; and how their designers endorse several narratives for an architecture of community and demonstrate that the sum of many more modest and smaller community inputs together can equally add value to the collective promise of a physical and tangible environment.

详谈社区建筑设计

与传统的设想相反，城市、城镇和居民区的公共空间不减反增，而且对于设计适应公共生活的建筑的展望也没有减少，而是日渐增加。在传统的城市中，公共空间是文化构建和深受欢迎的政治实践活动的关键场所，所以很少有或者根本没有其他进行公众集会的地点，或是用于参与活动、聚会和交际的场所。然而，如今共享空间日益增加，分布于各处，位于城市中央的公共空间是唯一的文化和政治活动场所的那段时期已经远去。可以肯定地说，公众集会和发表公开言论的场所大大增加，其中包含一些充满活力的新型公共场所和空间。在本章节中提到的图书馆、剧院、宗教场所、体育设施、社区中心、社区公园和城市基础设施，都是当代人的会面和交际场所的缩影。面对这种实用的"共享"空间的不断增加，我们该如何定义其价值和意义？

在城市化扩张、公共空间的用途多元化和文化表达场所日益增加的时代，期盼所有新型公共空间能够实现传统公共空间和参政议政空间的角色，似乎有些奇怪。那么新型空间的作用是什么？设计中所遭遇的挑战和关注点又引发了什么样的思考？

首先，城市公共空间不再是文化构建和政治观点构建的主要场所，而是这多姿多彩的世界的一个组成部分。一般而言，文化和社区不再仅由出现在城市空间和城市中特定的会面场所中的人群所构建。文化正越来越多地通过书籍、杂志、电视、音乐，尤其是基于数字空间和网络的新媒体等各种媒体形成。事实上，关于建成环境实用性的问题层出不穷，而且某些联合国教科文组织世界文化遗产在 Instagram 上比在现实中更受欢迎。这是否表明，我们想远离尘世喧嚣，进入一个瞬时、干净、充满正能量、美好的网络世界？与这种可能会撤离的情况相反，

Narratives for Community Architecture

Contrary to conventional assumptions, communal or public space in cities, towns and neighbourhoods is not in decline, but is instead expanding, and the prospect to orient architecture towards communities is not dwindling, but increasing. In "classical cities", public spaces were the key sites of cultural formation and popular political practice, with little or no existence of other sites for public gathering or places of participation, meeting and mingling. Today, however, the sites of shared space are plural and distributed, and we are far removed from the times when a city's central public spaces were the only cultural and political sites. The sites, where public gathering and expression occur, have most certainly proliferated to include the dynamics of an array of new types of places and spaces. The libraries, theaters, places of worship, sports facilities, community centers, neighbourhood parks and urban infrastructures portrayed within this section are snapshots of the many contemporary places used by people to meet and connect in physical space. Facing this increased availability of "shared" space, how do we define its value and meaning?

In an age of urban sprawl, multiple usages of public space and an increase of locations for cultural expression, it seems odd to expect that all these new forms of communal spaces would fulfil their traditional role as spaces of civic formation and political participation. So what could their new role be? And what are some of the challenges and concerns to reflect on?

To start with, urban public space is no longer the main site for civic and political formation, but one component in a multi-coloured field. Culture and community in general are no longer only formed in circuits of flows related to the urban or particular places of encounter within the city. Culture is increasingly formed through media such as books, magazines, television, music and especially new media built around digital space and online networks. In fact, questions are being raised about the relevance of the built environment, and

Galtzaraborda停车库项目，VAUMM
Galtzaraborda Parking, VAUMM

似乎有一种同样越来越强烈的愿望，那就是再次与我们自己产生连接，在现实生活中来一场相遇。特别是自从生活和工作越来越多地发生在虚拟空间之后，人们开始回到城市，在真实环境中参加活动、交流意见。现实和虚拟空间之间的这种分裂状态最近有了最新进展，那就是《精灵宝可梦 Go》的突然大获成功，这是一款将数字游戏叠加在现实世界之中的增强版现实游戏。这种增强版现实产品会一直新鲜有趣吗？或者这种趣味性会不会持续不断地涌现？这可能是以现实世界为基础的更广阔层面的游戏之始吗？

更多的关注表明"邂逅"场所、共享和"开放的"公共空间已经逐渐私有化，并且受到控制和监管。由于私有化、过度监管以及最近由于社会冲突激化和恐怖袭击事件增加所导致的公共空间军事化管理等因素正在逐渐侵蚀公共空间，也导致民众对此的焦虑不断上升。这些担忧和焦虑似乎伴随着一种趋势，那就是将"公共空间"的概念局限于公有制的传统户外空间。这种反应完全正确吗？公众联系和交流的机会在空间上真的如此受限吗？在本章节所呈现的五个项目图文并茂地介绍了人们在剧院中的聚会、在社区中心和咖啡馆的活动，并且在花园和城市电梯中创建了交流场所。也许并不是其所有制或外观使它们看起来像是"公共"场所的，而在于其公共用途，即不同的人会在这些地方举办不同的活动。如果以这种方式来考虑空间设计问题，并忽略其所有权和传统形式的话，那么在城市、城镇和社区中这些不断扩大的、往往是非正式的、分布于各处的公共空间很可能具有一种隐藏的潜力。

由精选的五个项目所展示的关于社区建筑的阐述可能不具备对于传统城市公共空间的传统影响力，也没有公民"力量"，但是这些项目促使人们保持谨慎的政治立场，减少冲动，这些效果无疑还是十分符

certain UNESCO World Heritage Sites, are "liked" more often on Instagram than in reality. Might this be times in which we want to step away from a crumbling reality to inhabit an instantaneous, clean, energy-rich and aesthetically pleasing augmented reality online? Contrary to this scenario of a possible withdrawal, there seems to be an equally growing desire to connect with ourselves again, and to meet physically. Especially since live and work increasingly take place in non-physical spheres, people are moving back into cities, meeting up for real events and exchanging views in physically tangible environments. This dichotomy between physical and digital space got a latest twist with the sudden success of Pokémon GO, the augmented reality game that superimposes a digital game over the fabric of the world. Will this flavour of augmented reality remain fresh, and will it keep emerging? Might this be the start of more extensive layers of play draped over the physical world?

Further concerns have been expressed that sites of "unpredictable encounter", communal and "open" public spaces, have been gradually privatised and made subject to control and surveillance. Anxieties are growing about the steady erosion of public space due to privatization, excessive policing and, more recently, the militarization of public space due to the rise of social conflict and terror attacks. These concerns and anxieties seem to come along with a tendency to confine notions of the "public" space to traditional outdoor spaces that are in public ownership. Is this reflex entirely correct? Are the opportunities for public association and exchange in space indeed so limited? The five projects presented within this section illustrate gatherings at theaters, activities in community facilities, cafés, and create places of exchange in gardens and urban lifts. Maybe it is not the ownership of places or their appearance that makes them "public", but their shared use for a diverse choice of activities by a range of different people. If considered this way, there might be a hidden potential in this expanding array of communal and shared, often informal and scattered spaces in cities, towns and neighbourhoods, regardless of their ownership or less

CKK Jordanki多功能音乐厅, Menis Arquitectos
CKK Jordanki Multifunctional Concert Hall, Menis Arquitectos

合公众的期望的。尽管传统公共空间被商业侵蚀，受到各种监督控制，但是各种分散的新型社区空间模式在面对当今经济、社会和生态危机时仍能提供强有力的对策。这些关于社区建筑的阐述可促成各种试验空间以及新的构想，进行尝试和经历失败，并且很可能创造新的公共理念。建筑阐述可以成为联系社区的强有力工具，也可以帮助我们探索目前急需的公共领域的意义与性质的转变。建筑阐述将帮助建立替代公共空间的基调与氛围、空间美学和物理架构，能巧妙地重新定义公众的社会生活形式，并改变城市公共文化的目的。建筑阐述可以带来声望，提供共享的经验、象征和层次丰富的背景。

关于可达性和用途多样化的阐述

可达性是许多社区项目成功的重要因素，通过增加参与公共活动的机会，更重要的是将不同的群体（不同社会阶层、文化水平、年龄段等）融合到一个有支持作用的、可以共享的大环境中。当这些体验通过多样化的用途得到重复和激励时，公共空间就变成了带有积极的公共意义的容器。有效的公共建筑为人们提供了见面和接触各种际遇的机会。这些会面通常都是偶然的，当然也可以是有人主动组织的。

由 Reiulf Ramstad、Lusparken 和 JST 设计的斯彻达尔文化中心通过开发艺术、舞蹈、音乐、电影和其他媒体等广阔的文化领域来支持其可达性的实现。该项目由图书馆、青少年活动中心、剧院、音乐厅、电影院以及可举办舞蹈、音乐、媒体、艺术和其他活动的场所组成。除此之外，同一屋檐下还包括了一座社区教堂和若干办公空间。该中心坚持，不只关注单一文化，而是鼓励各种不同的团体共同分享公共空间，举办各种活动。

conventional appearance?
The narratives of community architecture displayed by the five selected projects might not have the traditional impact or civic "power" of the classical city's public space, but the politically modest and smaller impulses generated by these projects are nonetheless still full of collective promise. Whilst the traditional public space is hijacked by commerce and surveillance, a model of manifold dispersed new community spaces could even provide a powerful counterpart in the face of the current economic, social, and ecological crises. These narratives for community architecture enable spaces for experiments and new imagination, attempts and failures, which could possibly create fresh philosophies of commons. Built narratives can become a powerful tool in uniting communities, and could help us explore a much needed transfer of meaning and nature of the public sphere. They will help establish the tone and mood for alternative communal spaces, their aesthetics and physical architecture, subtly redefine performances of social life in public and transform the purpose of urban public culture. They can provide a reputation, shared experiences, symbols and a rich and layered context.

Narratives of Accessibility and Versatile Rhythms

Accessibility is paramount to the success of many community projects, both by means of increasing opportunities to participate in communal activity, and even more importantly by means of bringing different (social, cultural, age, etc.) groups together in a supportive context of mutual enjoyment. As these experiences are repeated and stimulated through a versatile rhythm of use, public spaces grow into vessels that carry positive communal meanings. Effective community architecture provides opportunities for people to meet and to be exposed to a variety of encounters. These meetings often take place by chance, but they can also arise out of active organizing.
The Stjørdal Cultural Center designed by Reiulf Ramstad, Lusparken and JST encourages accessibility by tapping into a broad cultural field, including a wide range of art, dance,

阿尔法多媒体图书馆，Loci Anima
Alpha Multimedia Library, Loci Anima

VAUMM 设计的西班牙 Galtzaraborda 停车库项目将以下两种阐述结合起来：这个基础设施连接了坐落在山坡上的几个居民区和社区便利设施，而在此之前居民楼之间都是荒凉的空地。现在这里能适应多种通行节奏的需求，从悠闲漫步的人到脚步匆忙慌乱的行人，都能照顾到。这两个项目能够促进公共空间中的人们有建设性的互动，这是一种在许多郊区社区已经被遗忘和忽视许久的推动力。

关于材料的阐述

对于波兰托伦的 CKK Jordanki 多功能音乐厅，建筑师费尔南多·曼尼斯采用一种当代的方式对传统的砖块进行了重新解读（托伦老城的建筑立面一般都采用这种砖，它受到了联合国教科文组织的保护）。建筑师小心翼翼地通过建筑材料构建了一种阐述，通过红与白之间的色彩对比凸显了新建的现代建筑和原有历史建筑之间的差异性。砖结构是对“托伦历史的一种赞美”，曼尼斯说，“但这也证明了，在尊重历史的前提下加以创新是可以做到的。”用于整座建筑中的磨损骨料很快就能在使用者和建筑物以及活动之间建立一种熟悉感，并创建彼此之间的文化纽带。

Loci Anima 的建筑师在法国的阿尔法多媒体图书馆中选用了一种充满变化的立面，并对天体结构的彩色材料进行了阐述，这些天体结构与建筑的不同部分、立面相关。

关于这些项目的阐述可以追溯到公共场合的总体活力，使用者、建筑物和活动的结合，共享的实体空间内的多种用途和需要的结合。建筑师和设计师通过实体材料（混凝土、金属和砖等）对使用时间进行了安排。客户的使命与激情、建筑功能、场地现状及其历史位置，以

music, film, and other media. Its facilities consist of a library, a youth center, a theater and concert halls, cinemas, and venues for dance, music, media, art and other activities. In addition to these it incorporates a community church and office spaces within the same envelope. The center resolutely moves beyond a focus on mono-culture, encouraging diverse groups and activities to share common spaces.

Galtzaraborda Parking in Spain, designed by VAUMM, combines both narratives: the infrastructure links several neighbourhoods located on a mountain slope to community amenities, where barren voids previously separated residences; and they accommodate for the rhythms of use and passage of multiple users in transit, ranging from the slow walk of some and the frenzied passage of others. Two projects work with the power of promoting constructive interaction among people in communal space, an impulse that has unfortunately been forgotten or overlooked in many suburban communities.

Material Narratives

For the CKK Jordanki Multifunctional Concert Hall in Poland, Fernando Menis uses a contemporary reinterpretation of the traditional brickwork that is typical for the facades of Torún's old town and protected by the UNESCO. The architect carefully constructed a narrative through materials and opted for an interplay of colors between red and white to accentuate the dichotomy between the added modern and existing historic architecture. The brickwork is a “tribute to the history of Torún”, says Menis, “but demonstrates that it is possible to be innovative while respecting the past.” The abraded aggregate applied throughout the building builds upon an immediate familiarity and establishes a cultural link between bodies (users), mass and matter.

The architects of Loci Anima chose to apply a variation of facades to the Alpha Multimedia Library in France, and to add a narrative based on the colour-material of the celestial bodies that are associated to the different parts and facades.

The narratives of these projects can be traced back to the

El Roure社区中心和La Ginesta图书馆,
Calderon–Folch–Sarsanedas Arquitectes
El Roure Community Center and La Ginesta Library,
Calderon-Folch-Sarsanedas Arquitectes

上因素均融于简单的材料阐述中，进而发展成一个完整的故事，留下了一种感觉强烈、情感充沛的效果。

关于透明和开放性的阐述

这些精选项目将渗透性作为很重要的元素，从而使社区设计成为一个整体。项目在改变人们对室内外空间的认知和感觉的基础上展开叙述。设计积极推动渗透性强的、有边界的共享空间，这些边界起到的是过滤器的作用而非实体界墙，或者是允许人们及其视线穿过的完全开放的区域。这些项目亦传输了这样的态度：激励室内外的使用者将自己身边的物质世界看成是共享和公共空间。

在斯彻达尔文化中心的一层，落地式玻璃窗在室内与室外建立了一种开放的交流模式，在公众区域和半公开区域、城市和建筑之间建立了透明度。该建筑运用对透明的诠释除去了建筑内部的实际障碍，也消除了使用者的心理障碍，增强其包容性。在 Calderon–Folch–Sarsanedas 建筑师事务所设计的 El Roure 社区中心和 La Ginesta 图书馆中，我们可以将项目中对透明度和开放性的诠释与完全融入其周围环境的建筑区分开来。该中心是社交活动和文化的催化剂，通过恢复周围河流的景色并与这些景色产生共鸣，从而与自然景色产生一种互补效果。

关于安全性和舒适度的阐述

社区空间的创立通常与营造安全、舒适的空间紧密相连：这些空间让我们感受到的不仅是身体方面的安全感，还有精神层面和社会层面的安全感，这种安全感和生活中其他的必需品一样重要。这些阐述

total dynamic of the public setting, the bringing together of bodies, mass and matter, and of many uses and needs in a shared physical space. The architects and designers apply a placement of time through materialization (in concrete, metal, bricks, etc.). The clients' mission and passion, the building program or function, its site context, and often its place in history are edited into a simple material narrative and then expanded into a complete story leaving a strong sensory, affective result.

Narratives on Transparency, Openness

These selected projects hold permeability as a very important element to bring the design for community together. They create narratives based on changing the perceptions and sensations of inside and outside space. They actively promote a porous, shared space with boundaries that function as filters rather than solid confines, or openings that permit people, views and flows to pass through or be perceived through. They instil narratives that stimulate the users within or outside the building to perceive the immediate physical world around themselves as shared and communal.

At the ground level of The Stjørdal Cultural Center, floor-to-ceiling windows establish an open communication between the interior and the exterior, a transparency between the public area and the semi-public area, between the city and the building. It employs a narrative of transparency to remove physical and mental barriers and amplify its inclusive character. In the El Roure Community Center and La Ginesta Library by Calderon-Folch-Sarsanedas Arquitectes, we can distinguish a narrative of transparency and openness at work, with the building being fully integrated into its surroundings. The center is a social and cultural catalyst which supplements a stretch of nature by means of recovering and resonating with the surrounding river landscape.

Narratives for Safety and Comfort

The creation of community spaces is often closely linked to the creation of safe and comfortable spaces: spaces that make us feel not only physically, but also mentally and socially safe,

作家剧院，Studio Gang Architects
Writers Theatre, Studio Gang Architects

提到了这样一种建筑，这种建筑允许人们在与他人合作，可能甚至是对抗的过程中，可以采取适当、舒服、安全和有益的方式进行。在许多著名的社区中发现了这样一个关键因素，就是在面对陌生人、不熟悉的事物或是突发事件之时，对于空间的磋商、人与人之间的交涉，人们基本上没有觉得受到威胁。Studio Gang 建筑师事务所设计的作家剧院就包含这种磋商，由此建成了一座给社区日常生活带来活力的建筑，也成为一个激动人心的区域文化场所。这座建筑的组织方式就像由不同的体量围绕着一个中心枢纽组成了一个村落，其安排和布局均与周围城镇的特点遥相呼应。

El Roure 社区中心和 La Ginesta 图书馆的建筑师围绕着综合体的"内部广场"或是中央广场开发了该项目的功能需求。这样一来，建筑就具备了舒适感，可促进综合体内的各种人群之间的互动。同样，在阿尔法多媒体图书馆中，建筑师在巨大的功能空间内及各种功能空间之间提供了零散的休闲空间。这些舒适而零散分布的休闲空间，无论是在楼梯之间还是楼梯下，无论是在走廊旁边还是与露台和花园相结合，都为人们的各种碰面方式提供了安全的环境。

如本文所述，项目将注意力从传统公共空间转移到对社区空间及交流空间的崭新表现形式和阐述上。这些项目展示了从传统城市公众空间到丰富多彩的公共区域的转换。这些项目为广泛的公共或准公共空间提供了解决办法，并密切关注了学校、图书馆等地（此处仅举几例）的微政治环境。这些项目所产生的乐观主义应该被视为一种推动力，这种推动力有意识地反映了人们在社区发展中的作用，并提高了人们的公共意识。

are considered as important as many of life's other necessities. These narratives cater for an architecture that permits to act in the manner we feel is appropriate, comfortable, safe, and rewarding, in collaboration and perhaps even confrontation with others. A key ingredient, found in many notable community environments, is the negotiation of space and of bodies in a way that people feel largely unthreatened in the company of strangers and unfamiliar things and occurrences. The Writers Theater by Studio Gang Architects is a building offering this negotiation, and it produces an architecture that energizes the daily life of its community, and becomes an exciting, region-wide cultural destination. The building is organized alike a village-cluster of distinct volumes that surround a central hub, with its arrangement and layout resonating the character of the neighbouring downtown.

The architect of the El Roure Community Center and La Ginesta Library developed the program's functional requirements around an "inner square" or central agora within the complex. In doing so they provide the comfort to enhance citizen interaction between the diversity of people the complex is hosting. Similarly, within the Alpha Multimedia Library complex, the architects provided pockets of breakout spaces within and between the vast functional programs. Comfortable pockets of spaces in-between and underneath staircases, alongside corridors and integrated within terraces and gardens that offer safe environments for encounters of all sorts.

As demonstrated through this article, the projects shift the attention away from the traditional public space onto new expressions and narratives for community space and exchange. They demonstrate the transformation from traditional urban public space into a multi-colored field of communal spaces. They offer solutions for the widening range of public or quasi-public spaces and closely observe the micro-politics that are at work in places such as schools and libraries, to name a few. Their built optimism should be seen as an impulse to reflect consciously our role in community development and to increase our awareness of the commons. Tom Van Malderen

CKK Jordanki多功能音乐厅

Menis Arquitectos

JORDANKI

该项目在使用上有很大的灵活性。根据客户的要求，设计师原本只需要设计一个音乐厅，但是最终，在预算不变的情况下，设计师将原来单纯的音乐厅变成了一个可以承办各种演出和活动的复合型空间。首先，演出大厅的空间能够轻松承办各种活动。例如，可以通过移动墙壁、改变可轻松移动的座位的数量将两个空间合并成一个大厅，也可以同时举办各种不同的活动。其次，因为其天花板设计非常具有活力，只要加以有效调整，该建筑就可以满足交响音乐会、室内音乐会、戏剧表演、歌剧表演和电影放映的需要，可以满足剧院设计师提出的任何声学要求。最后，音乐厅可以朝向室外开放，将室内舞台与公园的露天表演结合起来。

设计师在整个建筑中都使用了皮卡多，这是一种混凝土和其他材料的混合物。CKK Jordanki 多功能音乐厅所使用的皮卡多，或者与从当地工厂（Ceramsus）回收利用的红砖混合以达到声音反射效果，或者与中国的红色火山岩混合以达到吸收声音的效果。皮卡多是由费尔南多·曼尼斯发明的一种创新技术，由混凝土和其他材料混合在一起，浇筑后分离。使用时除了有表面粗犷的效果之外，皮卡多还可达到极好的声学效果。曼尼斯首次把它使用在岩浆艺术与国会（位于加那利群岛的特纳里夫岛，2005 年）项目中，将混凝土与当地的火山石混合使用。在 CKK Jordanki 多功能音乐厅项目中，西班牙和波兰的建筑研究所分别对该技术进行了进一步研究、测试和证明。这个城市几乎所有建筑的外立面都使用红砖，CKK Jordanki 多功能音乐厅项目使用红砖，既是曼尼斯对这一传统材料的当代解读，同时显而易见也是在传承这一城市的文化遗产。

对礼堂空间及其外形的定义是通过一个始终与声学有关的交互式过程实现的，直至得出了这个最终外形。混凝土的可塑性使其能够以多种不同的方式用于音乐厅，因为设计师可以控制其几何形状（相当于液体石头），并按照模板浇筑合适的形状。这样设计师就可以控制听众听到的最初的声音反射效果了。此外，碎砖和混凝土混合体的表面处理给我们带来一种扩散的效果，这种效果是其他材料很难达到的。

礼堂的可活动部分的面积达 80 ~ 140m^2，因每个模块的重量不同，可活动部分的重量从 11t 到 20t 不等。每个模块都能够独立移动 3 ~ 5m 的高度，从而可以根据需要调整大厅的几何形状和体积：将体积

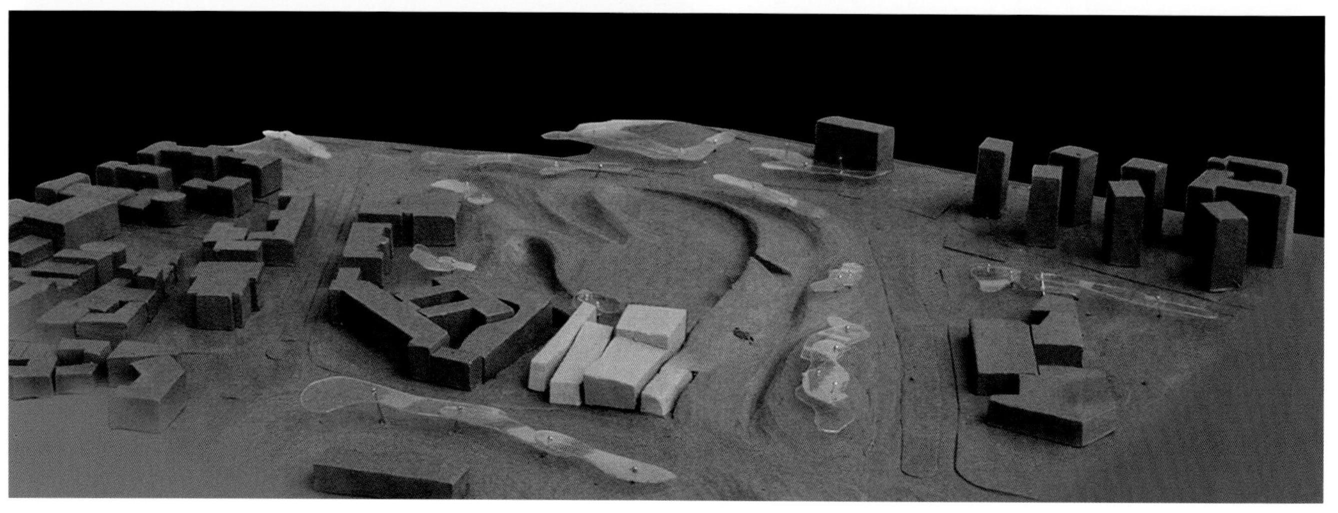

8200m³、混响时间为 1.85s 的空间变成体积 6800m³ 的空间，其混响时间也减少到 1.35s。通过增加额外的吸声模块，混响时间可以达到 1.2s。这样大厅几乎涵盖了所有演出活动的混响时间范围：混响时间 1.85s 的交响乐、1.6s 的歌剧和 1.2s 的戏剧。

CKK Jordanki Multifunctional Concert Hall

The program is characterized by a great flexibility at such extent that a building, which according to the client's brief, was meant to be only a concert hall, ended up being a space for all kinds of concerts and events, within the same initial budget. First of all, the theater space is able to adapt easily to different capacities. For instance you may join two rooms to act as the main theater by moving the walls and changing the number of the easily removable seats. It is possible to hold several separate and simultaneous events. Secondly, thanks to its dynamic ceiling, the building can be tuned to effectively absorb symphonic performances, chamber, theater, opera, and film and meet any acoustic requirements the theater designer requires. Finally, the concert hall can open to the outside, allowing to join the interior stage with the park outside for outdoor performances.

Picado, used in the entire building, is a mix of concrete and other materials. At CKK Jordanki, it was mixed either with reclaimed red bricks from a local factory (Ceramsus) for the sound reflection effect; either with a volcanic reddish stone from China, for its sound absorbtion effect. The picado is an innovative technique, conceived by Fernando Menis, consisting of mixing concrete with other materials and break it

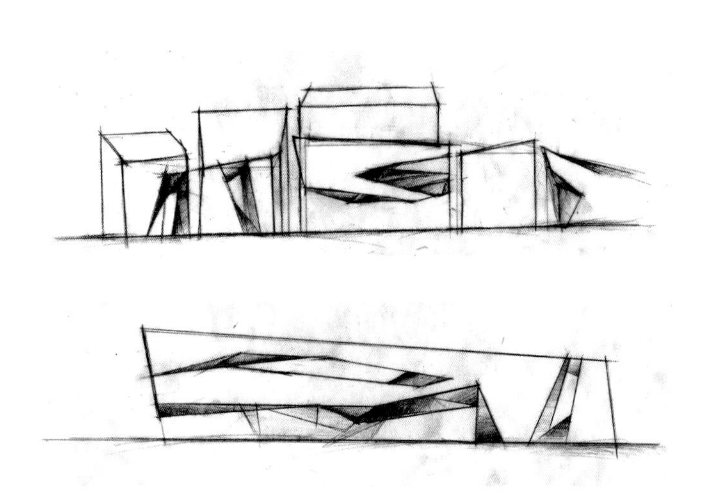

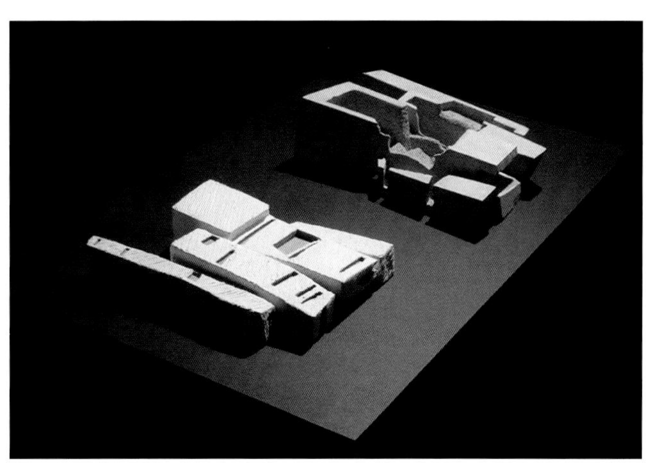

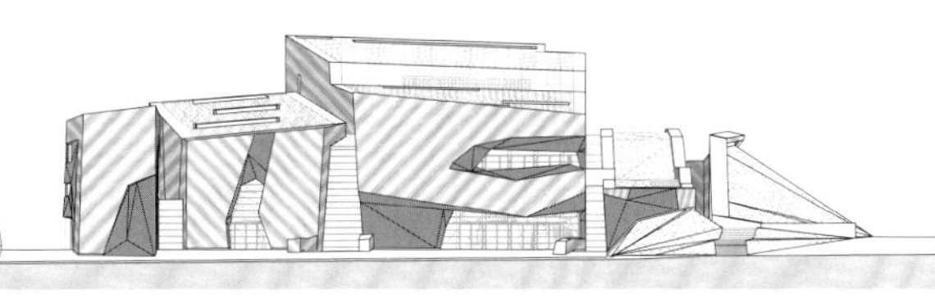

东立面 east elevation

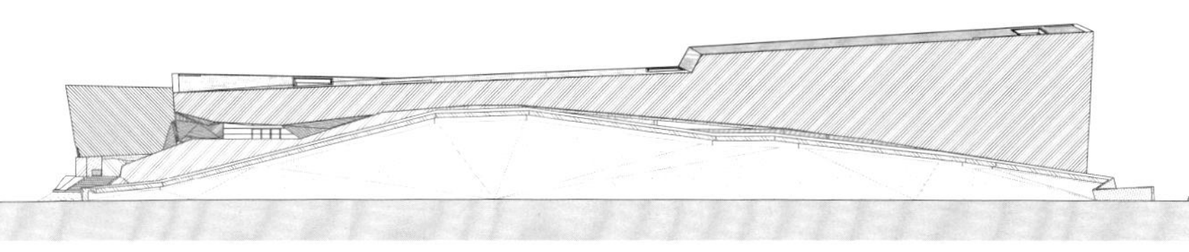

北立面 north elevation

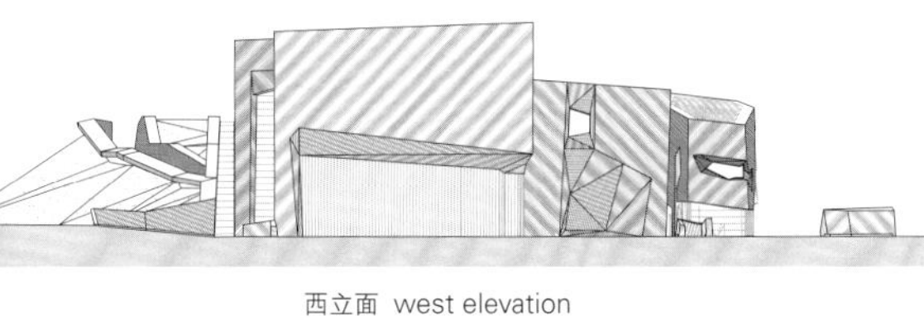

西立面 west elevation

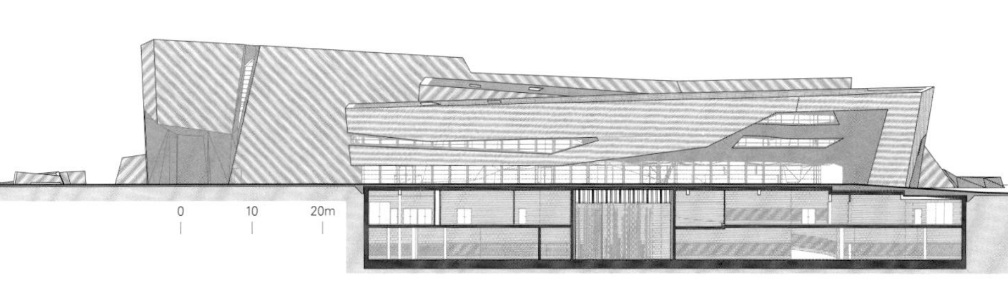

南立面 south elevation

CK JORDANKI

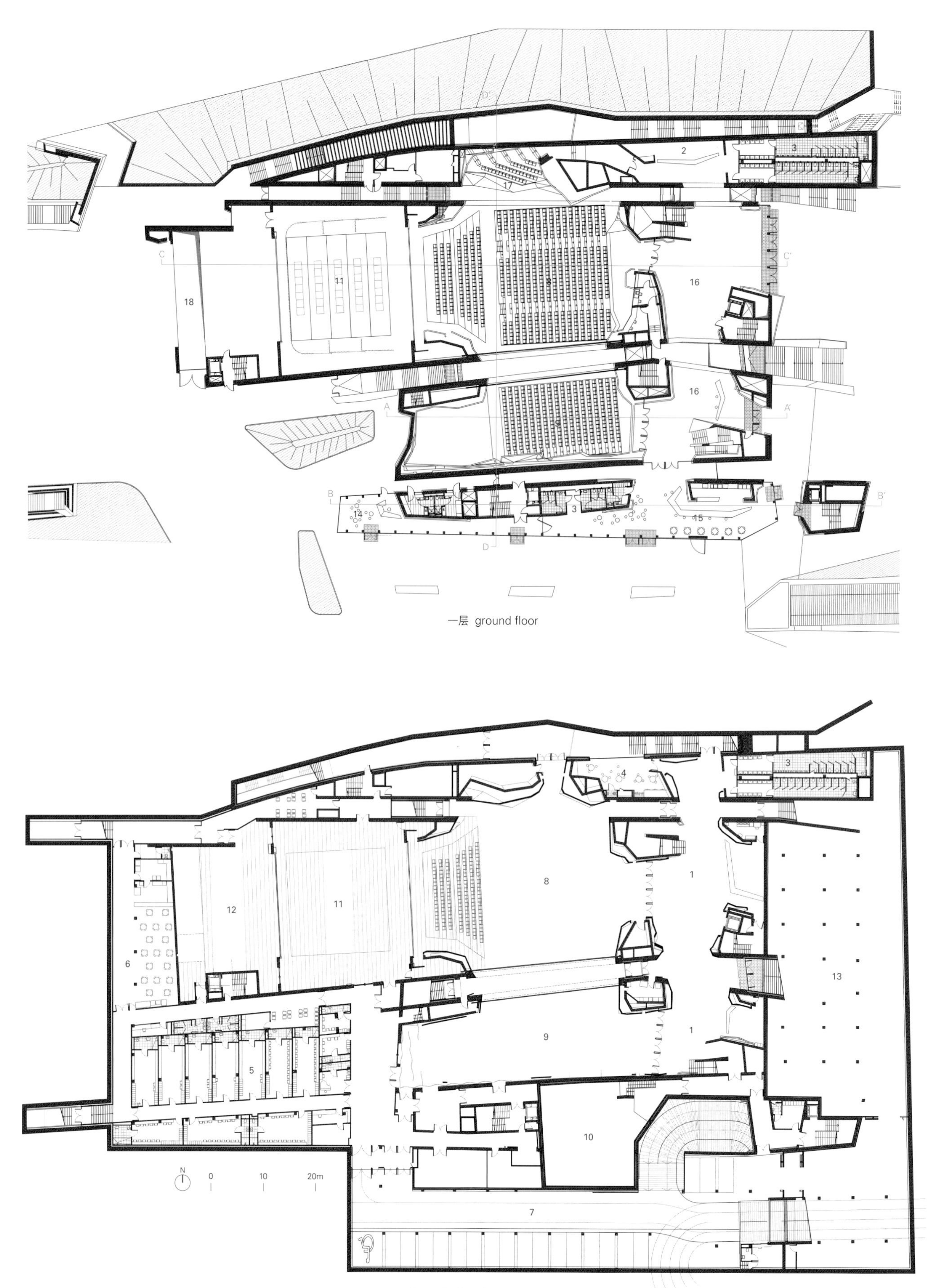

一层 ground floor
地下一层 first floor below ground

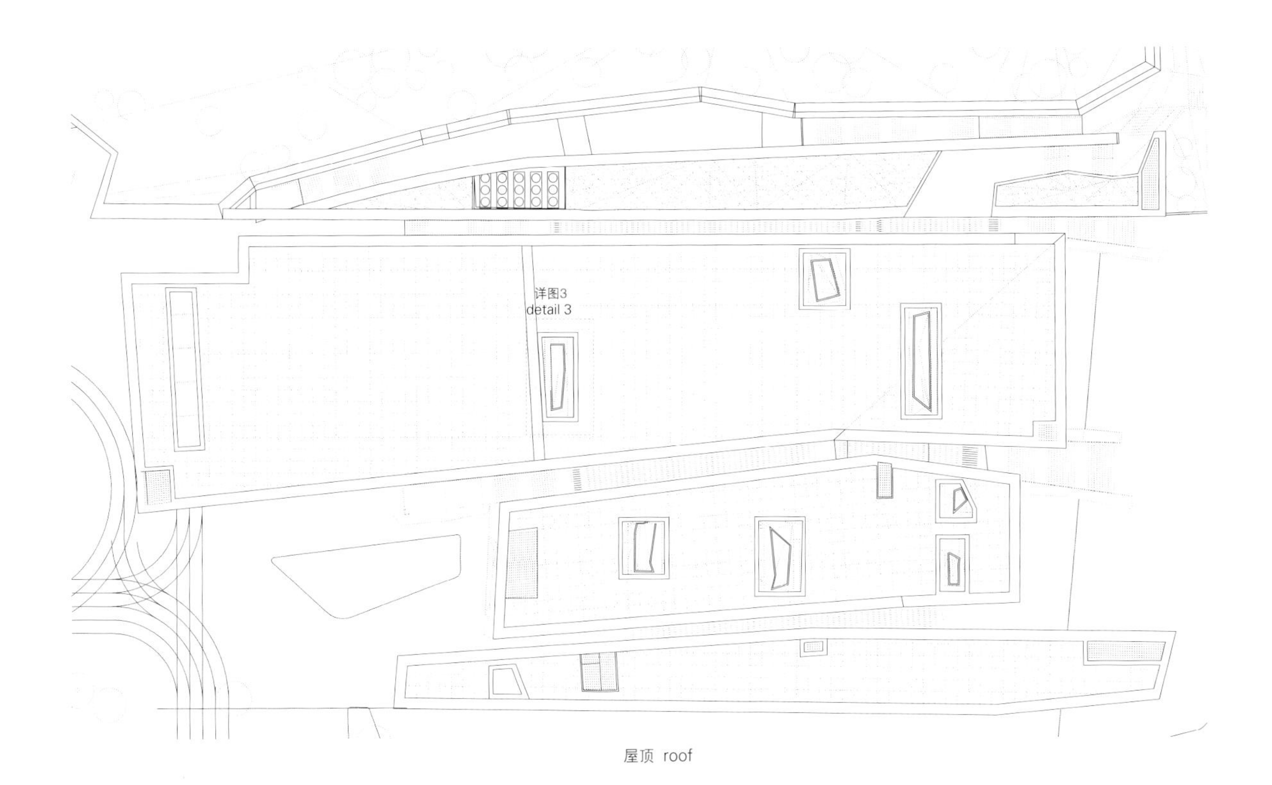

屋顶 roof

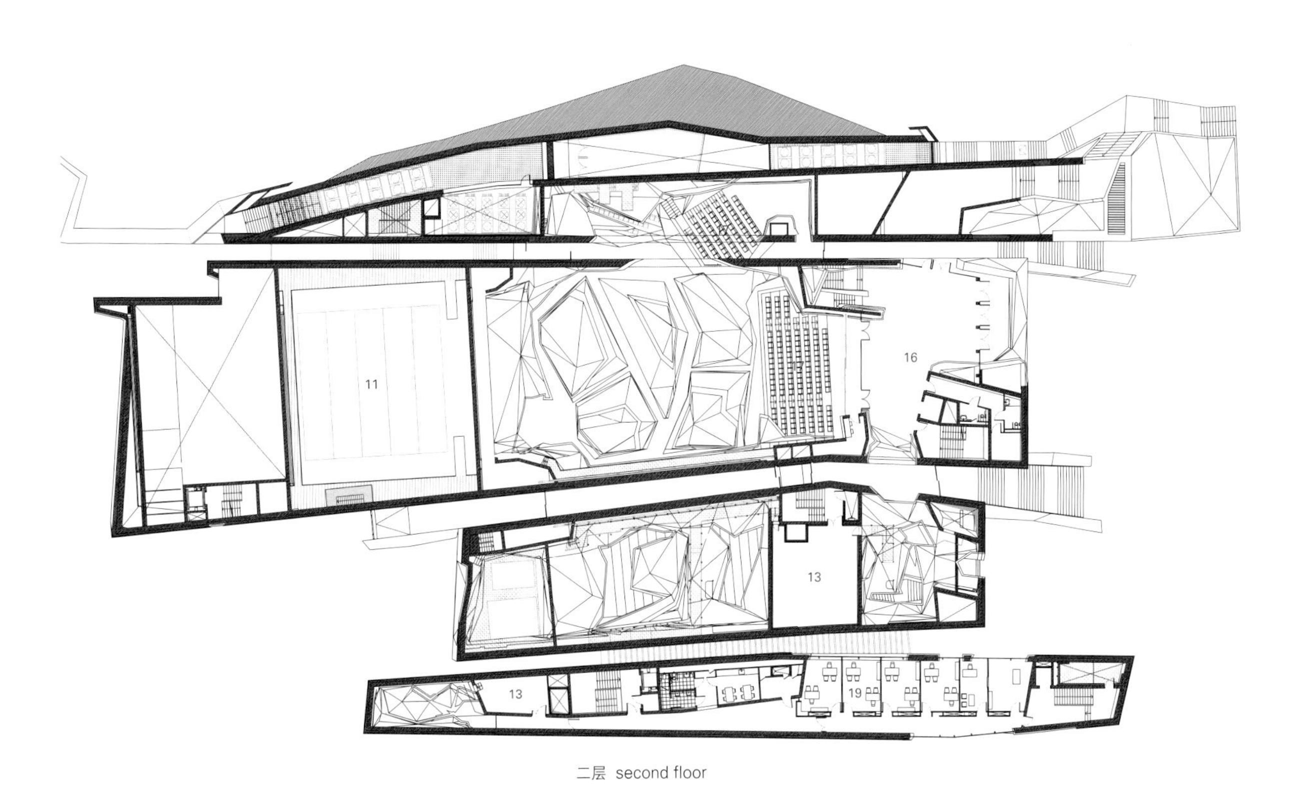

二层 second floor

1. 门厅 2. 衣橱 3. 卫生间 4. 母婴室 5. 更衣室 6. 餐厅 7. 停车场 8. 主要音乐大厅 9. 室内乐大厅 10. 排练室 11. 场景盒子
12. 储藏室 13. 技术设备间 14. 接待处 15. 自助餐厅 16. 大厅 17. 包厢 18. 室外舞台 19. 办公室
1. foyer 2. wardrobe 3.toilets 4. nursery 5. locker rooms 6. restaurant 7. parking 8. main concert hall 9. chamber hall 10. rehearsal room
11. scene box 12. storage 13. technical room 14. reception 15. cafeteria 16. hall 17. balcony 18. exterior stage 19. office

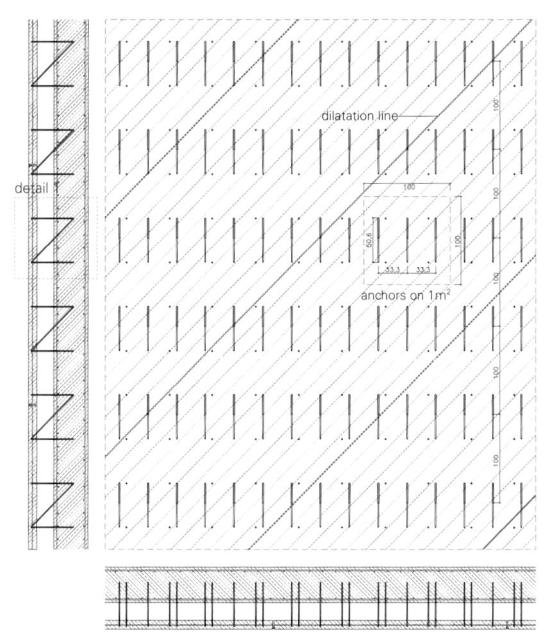

墙面固定系统
system of wall mounting

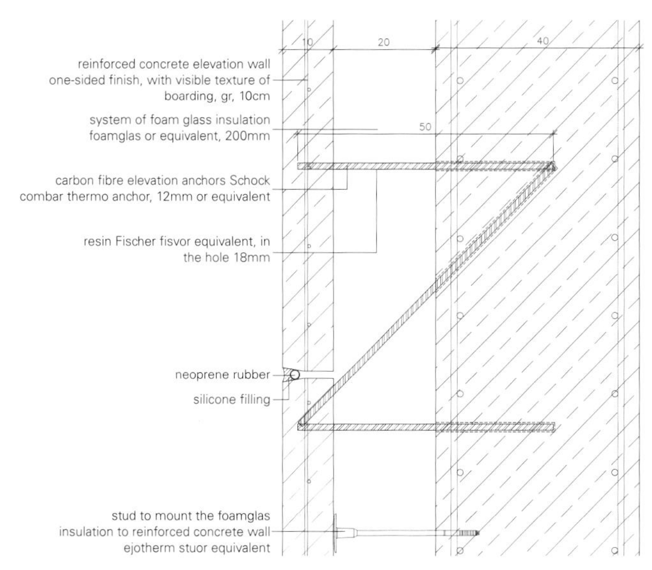

详图1 detail 1

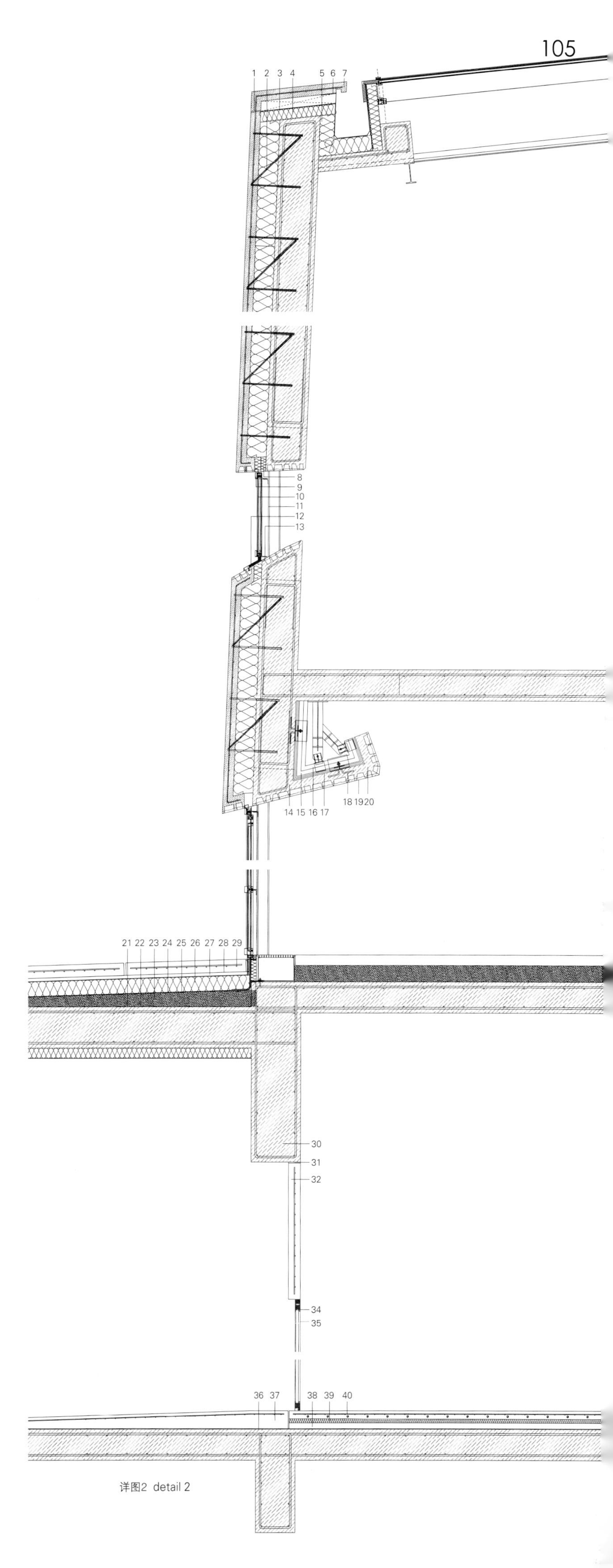

详图2 detail 2

1. overflow hole 2. carbon fibre elevation anchors Schock combat cfrp (or equivalent) 3. bituminous roofing felt x1 4. foamglass insulation - foamglas system (or equivalent),thk. 10 cm 5. mineral wool 6. metal sheet gutter 7. prefabricated concrete plates, thk. 8cm evening up concrete layer, 2-3cm 8. thermal insulation 9. steel window framing - Jansen (or equivalent) 10. elevation glass 4+16+4 - glaspol (or equivalent) 11. heb steel profile 12. thermal insulation with metal sheet coating on the one side and waterproof layer on the other 13. steel l-profile 60x60mm 14. steel bar connecting picado to the wall 15. metal element mounting picado 16. pipe profile mounting picado 17. supporting structure of picado 18. metal element mounting picado 19. metal mesh 20. conglomerate of brick and concrete, thk. approx. 12 cm 21. cross-linked polyethylene foam - polifoam (or equivalent) 22. geotextile delta biotop ts 30 (or equivalent) 23. sloping layer of concrete 24. bituminous roofing felt x1 25. geotextile delta biotop ts 30 (or equivalent) 26. extruded polystyrene - knauf, thk. 15cm (or equivalent) 27. geotextile delta biotop ts 30 (or equivalent) 28. sloping layer of concrete, 2-3cm 29. prefabricated reinforced concrete plates, 240x240cm, with a texture made with natural aggregate, 12cm 30. reinforced concrete wall 31. dilatation of reinforced concrete elements 32. reinforced concrete wall, thk. 12cm 34. door frame 35. door 36. cross-linked polyethylene foam - polifoam (or equivalent) 37.concrete flooring reinforced with fibers, thk. 18cm, polished 39. hard mineral wool panel - Stropoterm, Isover (or equivalent) 40. polished concrete flooring with floor heating, thk. 8cm

afterwards. Besides achieving a rough expression, the picado allows excellent acoustics results. The first time Menis used it, was for Magma Art & Congress (Tenerife, Canary Islands, 2005), by mixing concrete with local volcanic stones. For the CKK Jordanki, the technique has been researched further, tested and certified by the Spanish and the Polish Building Research Institute, respectively. The red brick is present almost on all the facades of the city and its use in CKK Jordanki is Menis's contemporary reinterpretation of this traditional material while a clear reference to the town's cultural heritage.

The definition of the auditorium space and of its shapes was made through an interactive process, always in relation to acoustics, until reaching the final shape. The plastic properties of concrete allows its use in concert halls in many different ways because it allows you to control the geometry (liquid stone) and adapt its shape to the formwork so that you can control the first sound reflections the listener received. In addition, the surface treatment of the crushed brick and concrete mix, allowed us a kind of diffusion, very difficult to achieve with other materials.

The moving parts of the auditorium have an area ranging from $80m^2$ to $140m^2$, having a weight that varies according to the piece, from 11t to 20t. Each of the pieces can move independently from 3m to 5m in height, allowing adjustment of the geometry and volume of the hall, depending on the needs of each moment, transforming a volume of $8,200m^3$ with a reverberation time of 1.85 seconds, into a volume of $6,800m^3$, with a reverberation time reduced down to 1.35 seconds. By adding additional absorption you could reach a reverberation time of 1.2 seconds, which covers the entire range of possible activities: 1.85 seconds for symphonic music, 1.6 seconds for opera and 1.2 seconds for theater.

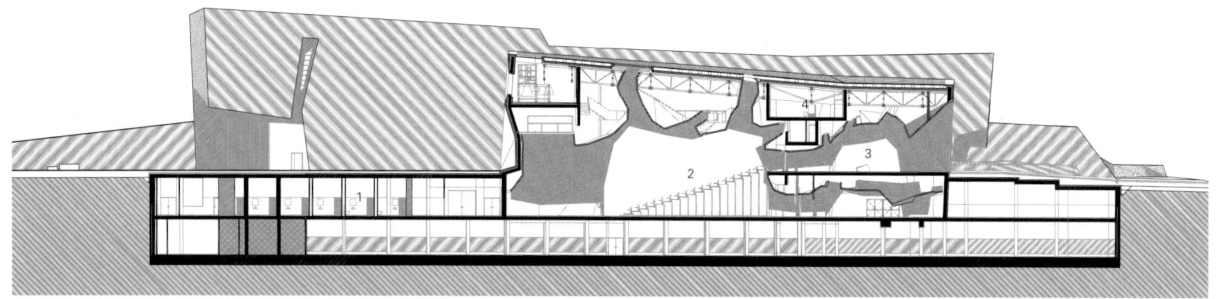

1. 更衣室 2. 室内乐大厅 3. 大厅 4. 技术设备间
1. locker rooms 2.chamber hall 3. hall 4. technical room
A-A' 剖面图 section A-A'

1. 接待处 2. 技术设备间 3. 排练室 4. 办公室 5. 自助餐厅
1. reception 2. technical room 3. rehearsal room 4. offices 5. cafeteria
B-B' 剖面图 section B-B'

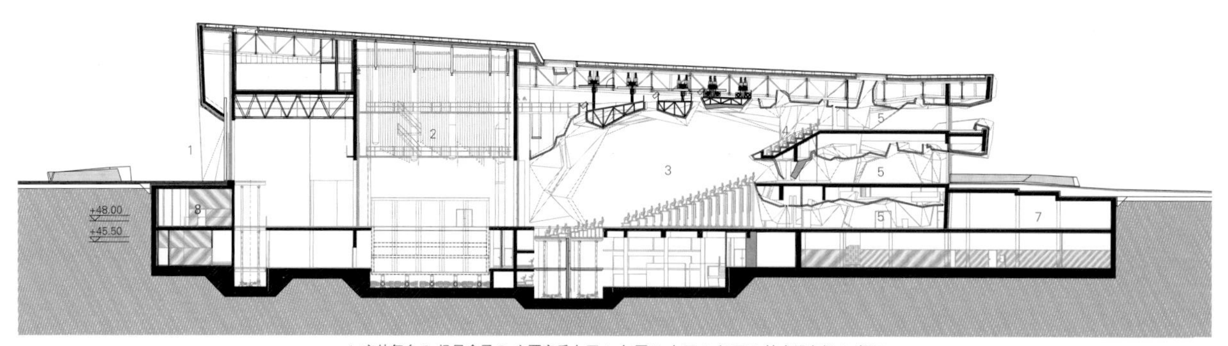

1. 室外舞台 2. 场景盒子 3. 主要音乐大厅 4. 包厢 5. 大厅 6. 门厅 7. 技术设备间 8. 餐厅
1. exterior stage 2. scene box 3. main concert hall 4. balcony 5. hall 6. foyer 7. technical room 8. restaurant
C-C' 剖面图 section C-C'

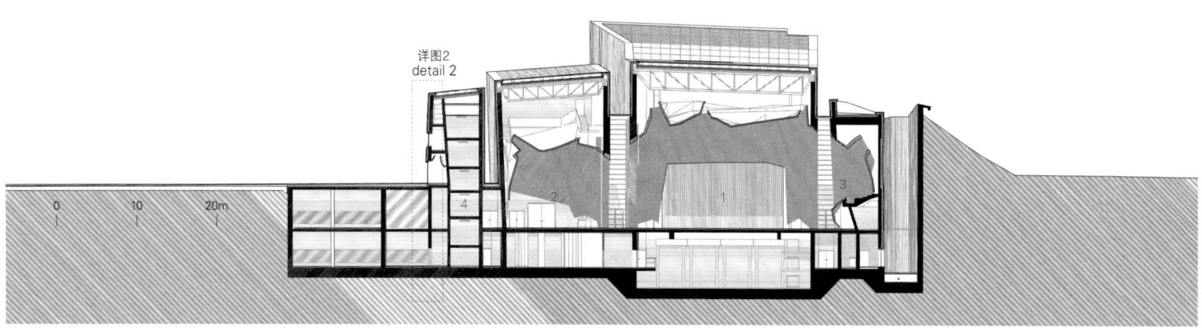

1. 主要音乐大厅 2. 室内乐大厅 3. 包厢 4. 卫生间
1. main concert hall 2. chamber hall 3. balcony 4. toilet
D-D' 剖面图 section D-D'

项目名称：CKK Jordanki / 地点：Torun, Poland / 建筑师：Fernando Menis / 合作建筑师：Karolina Mysiak, Jaume Cassanyer, Javier Espílez / 草图设计阶段合作方：José Antonio Franco (Martínez Segovia y asociados) for structure / José Luis Tamayo for stage equipment, Pedro Cerdá for acoustics / 波兰团队：Jacek Lenart (STUDIO A4 Spólka Projektowa z o.o.)_Supervising Architect / Tomasz Pulajew (FORT POLSKA Sp. z o.o.) for structure / Pedro Cerdá for acoustics / 设计阶段合作方（波兰）：ELSECO Sp. z o.o. for electricity; Iskierski Mariusz. Biuro Inzynierskie for HVAC Systems, Plumbing, Telecommunications; Pracownia Architektury i Urbanistyki SEMI for urban planning / 用途：Cultural, Concerts Hall / 建筑面积：21,837m² / 总建筑面积：46,971m² / 结构：reinforced concrete and steel / 材料：concrete and red bricks / 造价：51,025,731 EUR / 设计时间：2011 / 施工时间：2013—2015

摄影师：©Roland Halbe - p.96~97, p.100, p.101, p.104, p.106; ©Jakub Certowicz (courtesy of the architect) - p.109, p.114~115; ©Patryk Lewinski (courtesy of the architect) - p.107, p.110~111, p.112

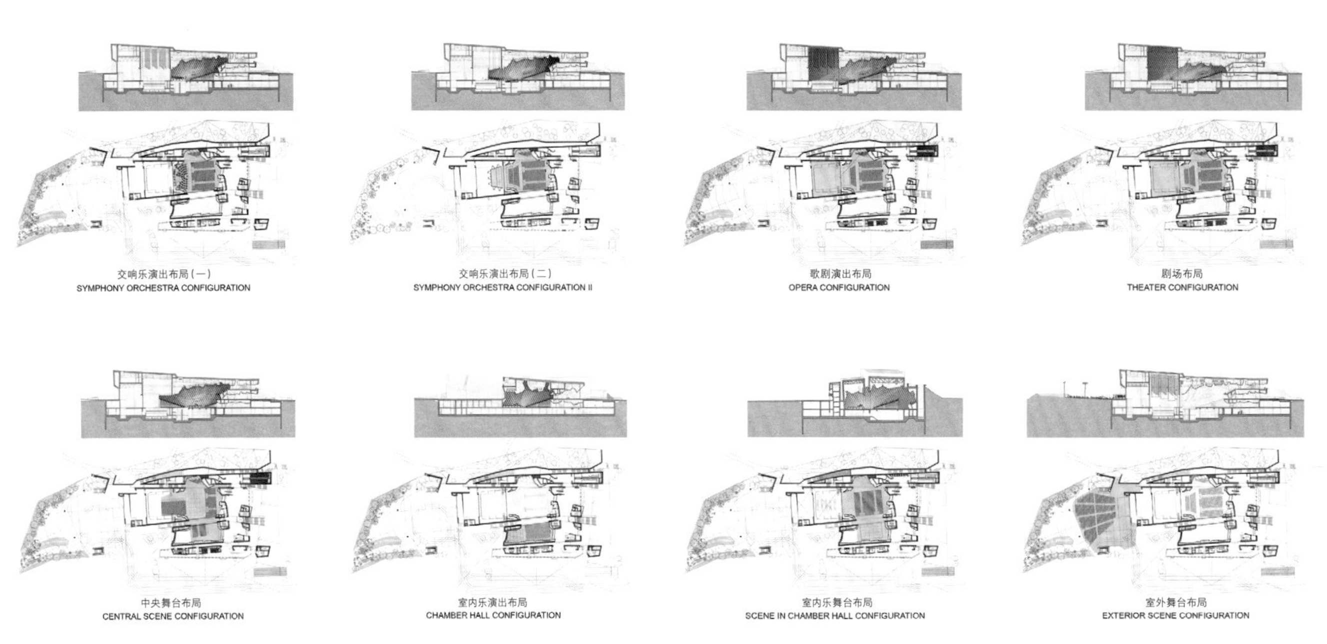

交响乐演出布局(一)
SYMPHONY ORCHESTRA CONFIGURATION

交响乐演出布局(二)
SYMPHONY ORCHESTRA CONFIGURATION II

歌剧演出布局
OPERA CONFIGURATION

剧场布局
THEATER CONFIGURATION

中央舞台布局
CENTRAL SCENE CONFIGURATION

室内乐演出布局
CHAMBER HALL CONFIGURATION

室内乐舞台布局
SCENE IN CHAMBER HALL CONFIGURATION

室外舞台布局
EXTERIOR SCENE CONFIGURATION

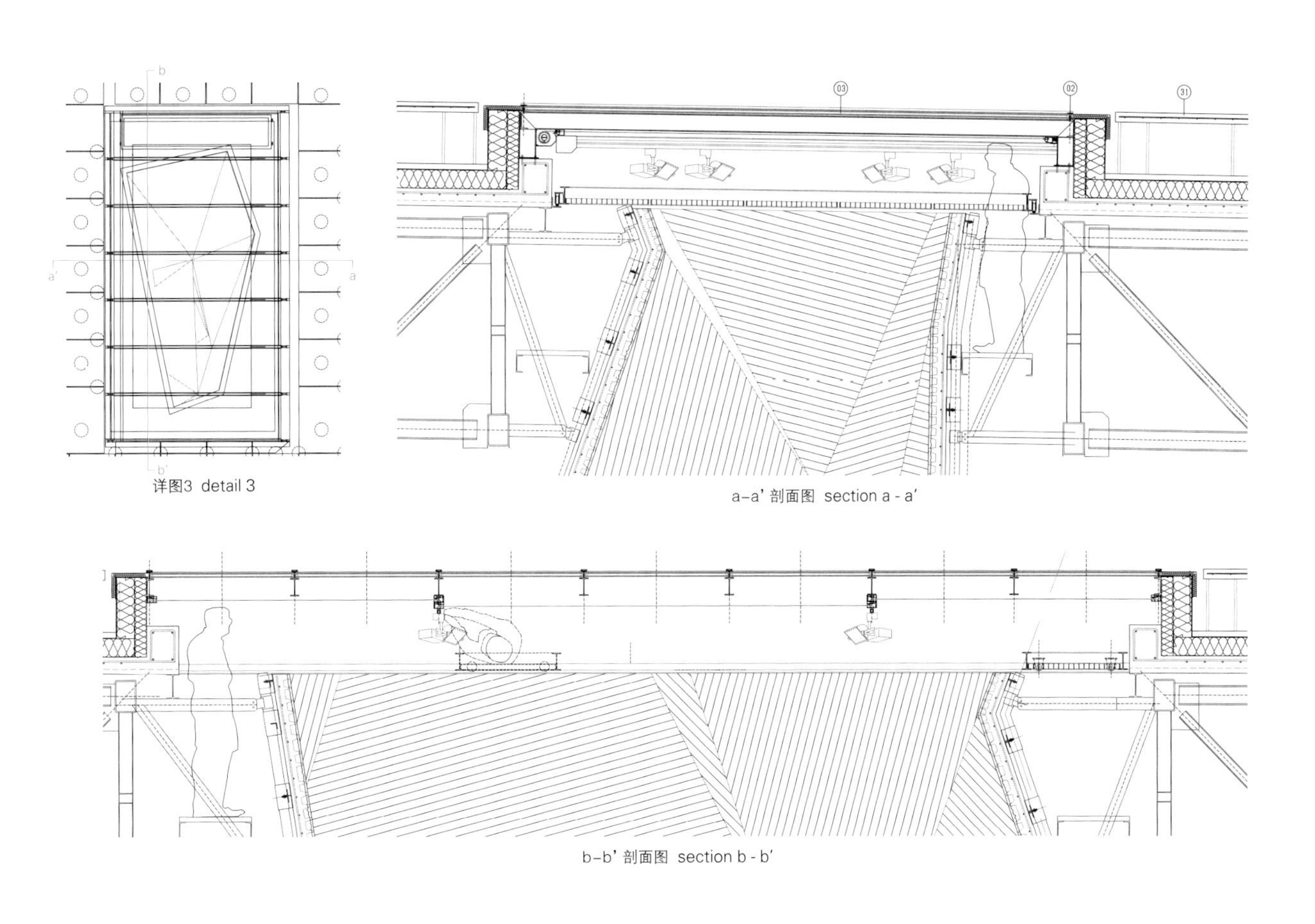

详图3 detail 3

a–a' 剖面图 section a - a'

b–b' 剖面图 section b - b'

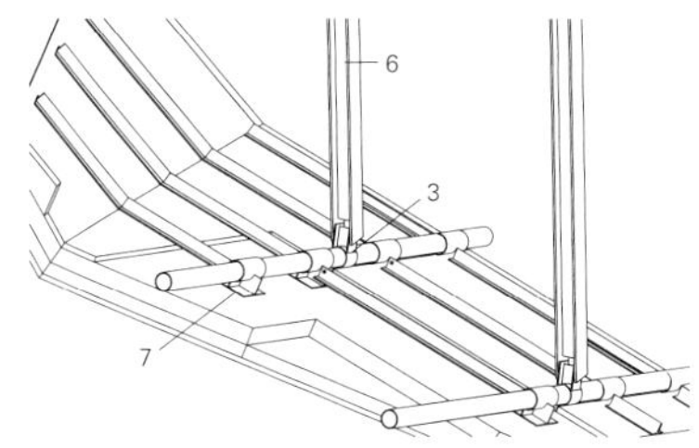

1. conglomerate of brick and concrete, total thickness approx. 14 cm, reinforced concrete with polypropylene fibers, with the addition of resins = 8cm, blend of concrete with broken clinker bricks +pp/hp, textured with pneumatic hammer from 6 cm to approx. 4cm
2. rebar of reinforced concrete wall
3. connecting element 1
4. steel pipe outer ø = 80 mm. thickness= 4mm
5. steel profile "I" 50x50x5mm
6. wide-flange I-beam heb 100
7. connecting element 2
8. silent block

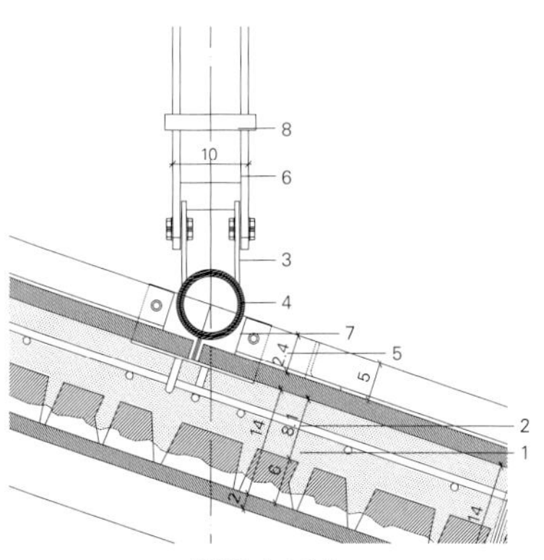

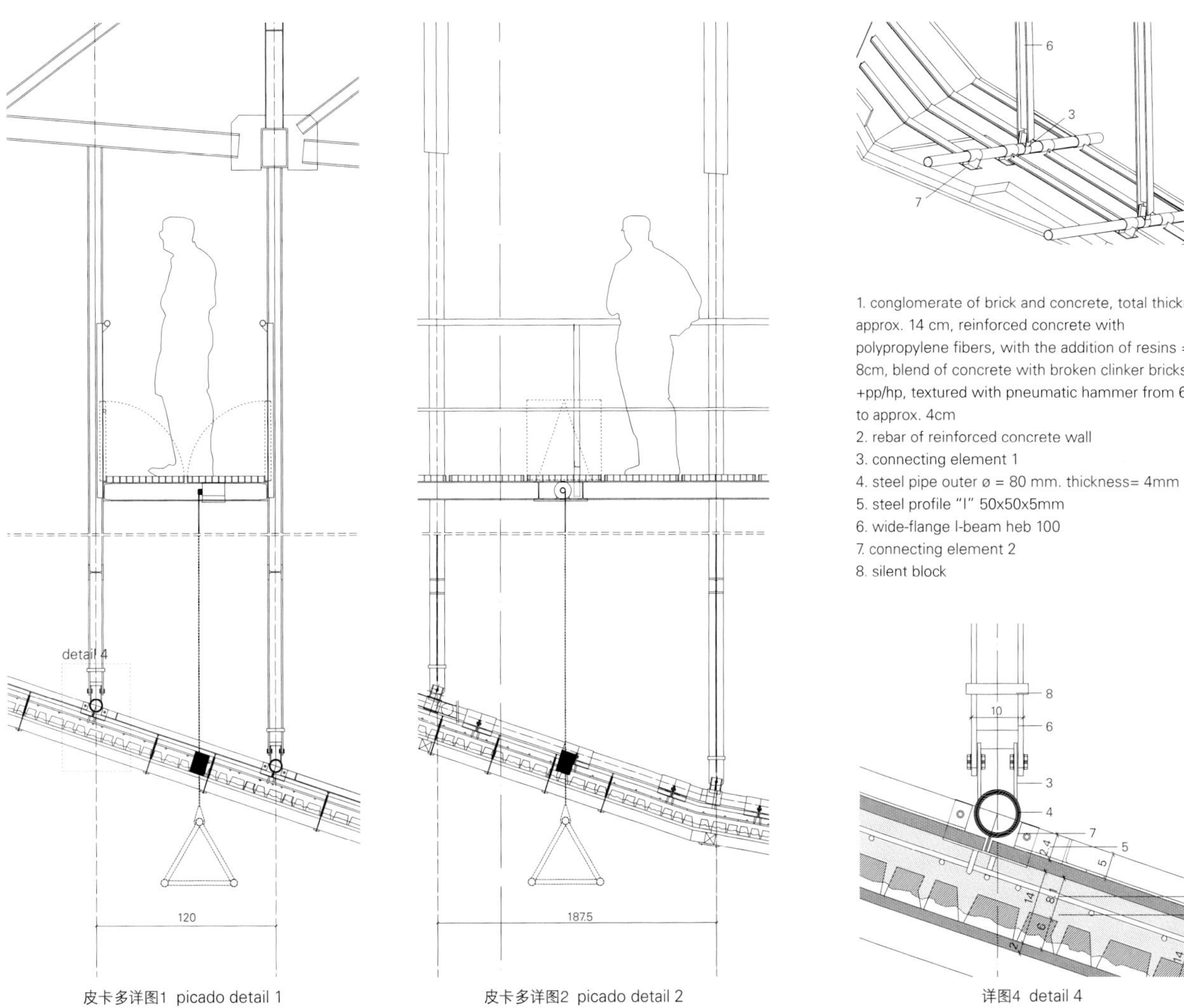

皮卡多详图1 picado detail 1

皮卡多详图2 picado detail 2

详图4 detail 4

ANGOULÊME

阿尔法多媒体图书馆

Loci Anima

在虚拟媒体的使用日益增加的背景之下，Loci Anima 建筑事务所设计了位于昂古莱姆的阿尔法多媒体图书馆项目。该项目位于昂古莱姆东北部，这里过去曾是工业区。该项目为目前正在进行彻底改造的 Houmeau 区增添了价值，改善了该地区市民的社会和文化生活。项目紧邻火车站，这意味着从这里可以轻松通往昂古莱姆的各个社区（同时，人行天桥也提供了直接通道），从而也能通向全世界。

项目设计的灵感源于斯堪的纳维亚风格，该建筑是新一代文化空间的实证，市民们在这里碰面、交流、学习、放松，这些都是著名的"第三空间"，保持中立并对所有人开放，不受任何年龄限制，这种空间类型旨在消除社会文化差异，淡化公共和私人、集体和个人之间的界限。"第三空间"是在20世纪80年代由城市社会学教授雷·奥尔登堡提出的，与"第一空间"，即家庭和"第二空间"，即工作场所截然不同。

怎样创造一个对所有人开放、不令人畏惧、有强大社会功能又让所有人都可以享受的空间呢？对于这个问题，Loci Anima 建筑事务所的创始人弗朗索瓦兹·雷诺用一些与童年和色彩相关的语汇做出了解答。她设计了一座有着"瑰丽风景"的多立面建筑，对着昂古莱姆的重要标志性建筑物，并且下至大地、上至天空都能相互凝望。多媒体图书馆由五块彩色的平行六面体组成，它们巧妙地互相堆放，从城市的高空望过去，阿尔法图书馆形成了一个 A 字。构成整座建筑的五个"世界（主题空间）"的每个部分都可扩展，由天体的彩色金属材料加以区分。"创造"世界是关于土星的，以煤灰色为主打色；"理解"世界使人想起月球和银色；"想象"世界代表木星，为青铜色；"从一个世界到另一个"世界代表太阳，为金黄色；最后，"制造"世界是红铜色，以此表达对火星的敬意。

对童年想象世界的含蓄表达也体现在设计和家具上，这些家具呈几何形状，又有趣、又实用，与多媒体图书馆所提供的多种不同用途相互呼应。环绕着每张桌子的木框架可让读者处于独立、专注的氛围中，同时也对外打开一扇窗。自然采光的大桌子可在小组活动中促进团队的交流。室内外座椅被设计成一些小物件的超大版，比如，做成超大版的安全别针或挂钩，坐起来有趣、亲切并可调节。建筑师在入口大厅设计了一段巨大的东向悬空楼梯，空出了大厅天花板下的空间，人们一走进建筑，就能产生一种每人都可拥有的专属空间感。另外，这些空间延伸到室外露台、花园，适合室内外使用，充分利用每处的自然光线，并根据温度和光照需求或喜好打开或关闭这些空间。据观察，城市居民越来越远离自然，而且渴望在生活和自然元素中找到丰富的梦想和快乐源泉，因此，Loci Anima 建筑事务所在此将这座建筑创造为一种生活的空间。阳光、空间、植物、水和金属都是这个项目建造过程中使用的材料。除了该建筑在环境方面实际的基本性能（低能耗建筑）之外，Loci Anima 建筑事务所独有的这种设计方式还赋予该项目一种象征意义和细腻感。

Alpha Multimedia Library

In a context of the increasing use of virtual media, Loci Anima has delivered the Angouleme Alpha Multimedia Library. Built on the site of a former industrial area in the north-east of the city, the Alpha Multimedia Library brings added value to the Houmeau district, which is being radically overhauled, improving the social and cultural life of people in this area. Its direct proximity to the railway station means that it is easily linked to all of the communes of Greater Angouleme (in time, a footbridge will provide direct access), but also to the world as a whole.

Inspired by the Scandinavian model, this building is concrete evidence of a new generation of cultural spaces where citizens are able to meet, to exchange, to learn, to relax, these famous "third place". Neutral and open to all, making no distinction in terms of age, this type of space seeks to erase socio-cultural differences and blur the boundaries between the public and the private, the collective and the individual. The "third place", an idea developed in the early 1980s by the urban sociology professor Ray Oldenburg, is different from the "first place", the home, and the "second place", the workplace.

How do you create a space that is open to all, that is never intimidating, that has a strong societal function and that everyone can own? In answering this question, Francoise Raynaud, founding architect of Loci Anima, uses vocabulary associated with childhood and with color. She then designs a building with multiple facades, a "rose of views", facing the important landmarks of Angouleme and that face each other from the earth and from the sky. Made up of five colored parallelepipeds, cleverly stacked one on top of the other, the multimedia library, viewed from the highpoint of the city, forms an A for Alpha and for Angouleme. Each of these five "worlds" which

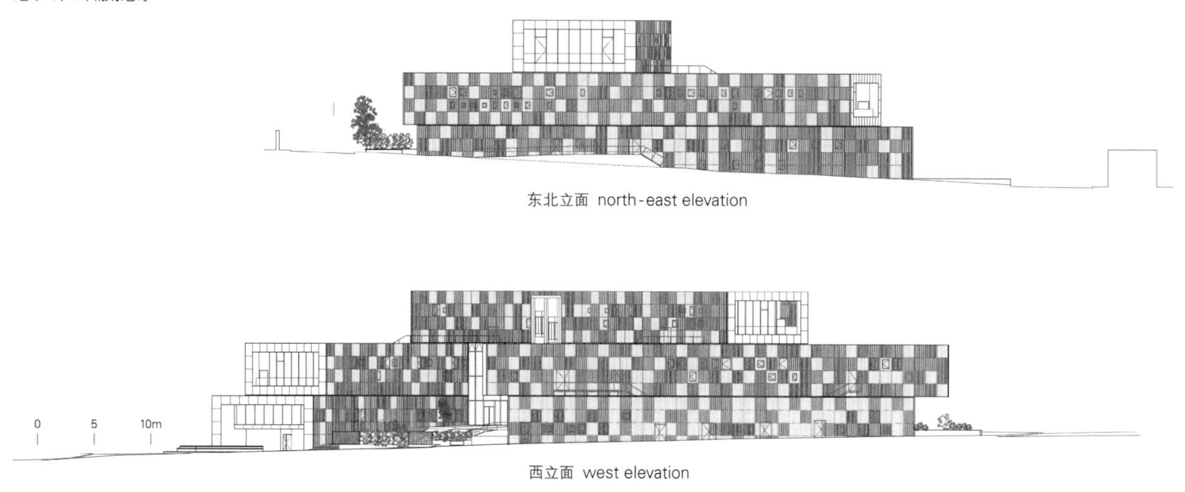

东北立面 north-east elevation

西立面 west elevation

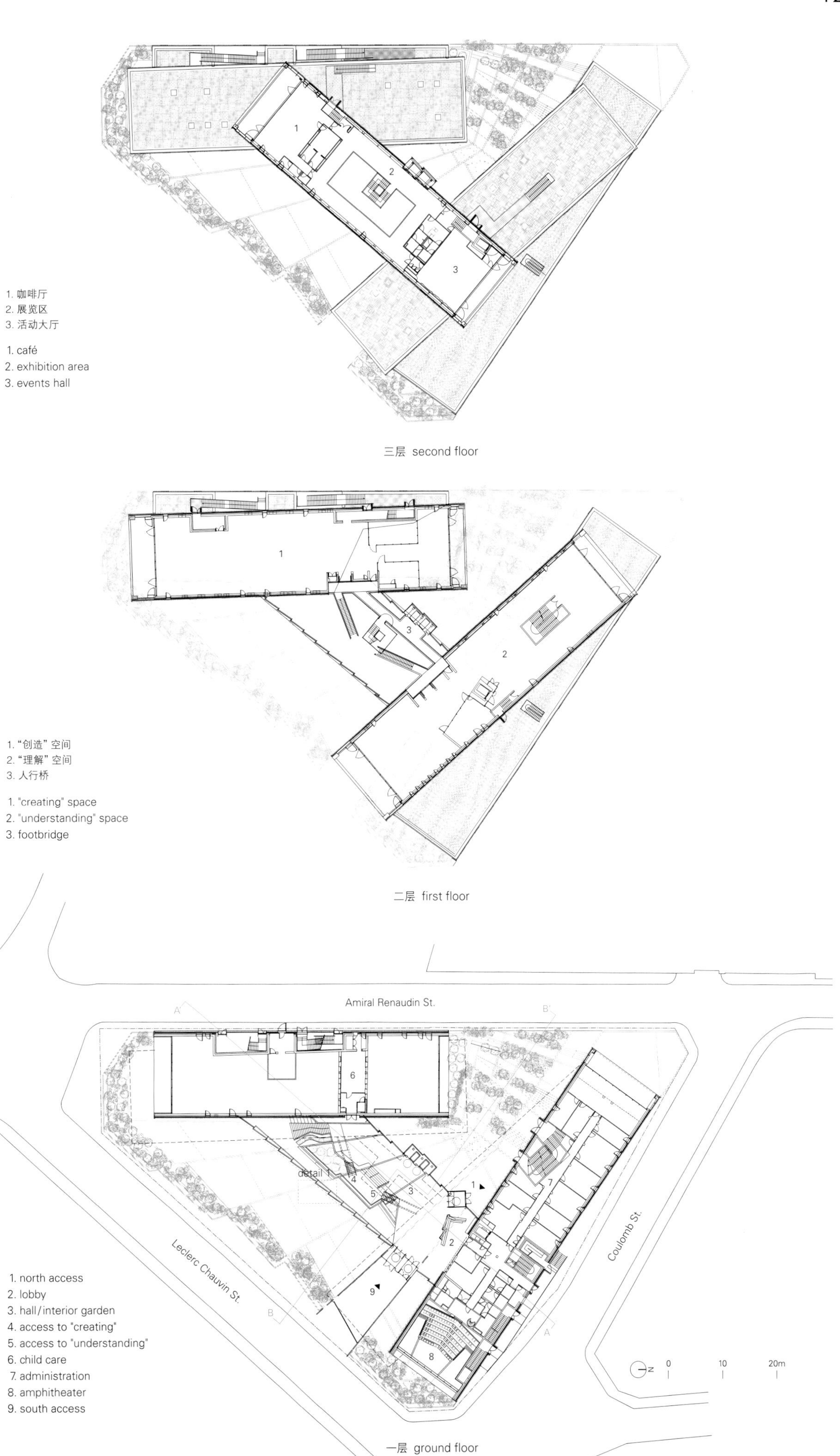
1. 咖啡厅
2. 展览区
3. 活动大厅
1. café
2. exhibition area
3. events hall
三层 second floor
1. "创造" 空间
2. "理解" 空间
3. 人行桥
1. "creating" space
2. "understanding" space
3. footbridge
二层 first floor
Amiral Renaudin St.
Leclerc Chauvin St.
Coulomb St.
detail 1
1. 北侧入口
2. 门厅
3. 大厅/内花园
4. "创造" 空间入口
5. "理解" 空间入口
6. 儿童看护室
7. 管理办公室
8. 半圆形露天剧场
9. 南侧入口
1. north access
2. lobby
3. hall/interior garden
4. access to "creating"
5. access to "understanding"
6. child care
7. administration
8. amphitheater
9. south access
0
10
20m
一层 ground floor

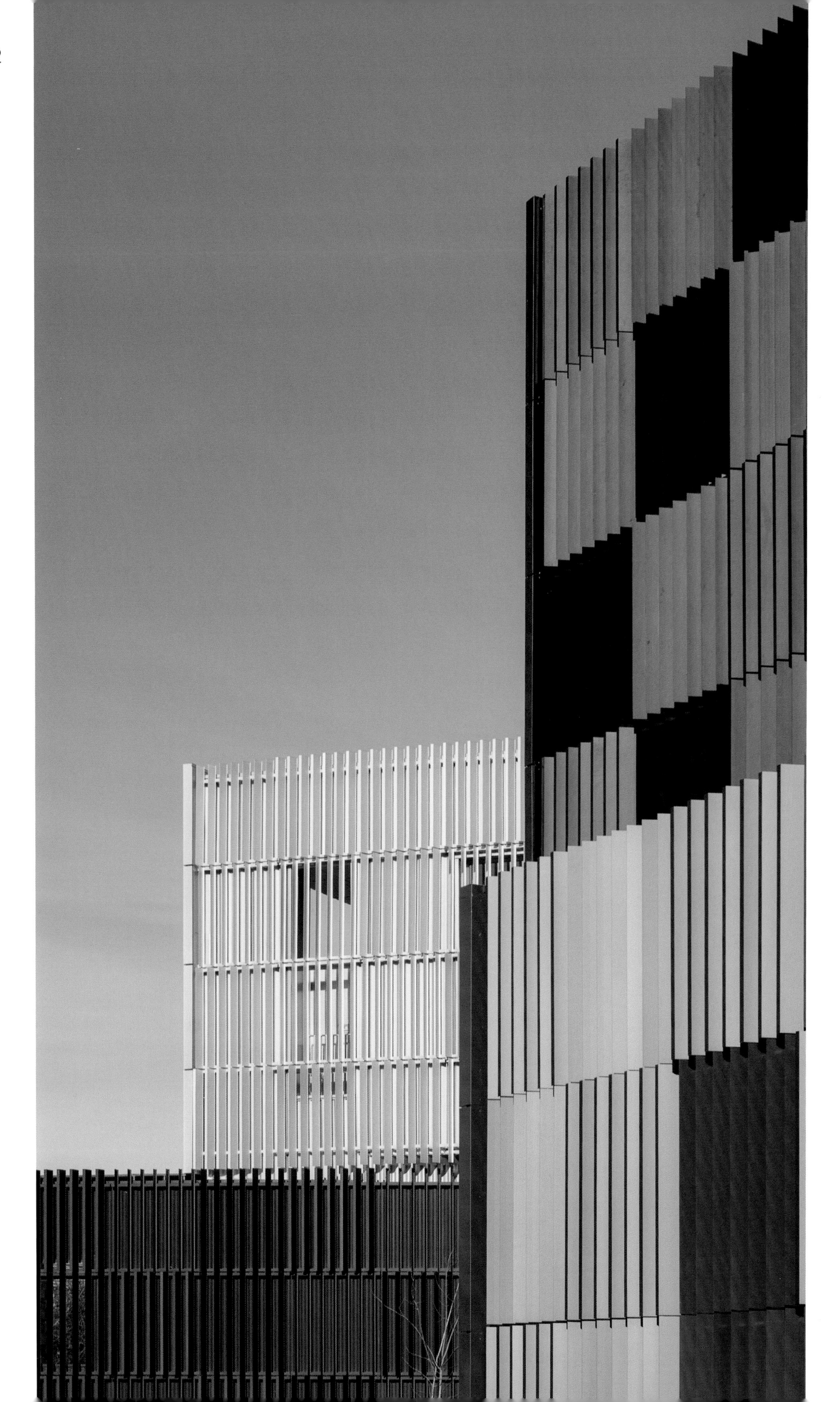

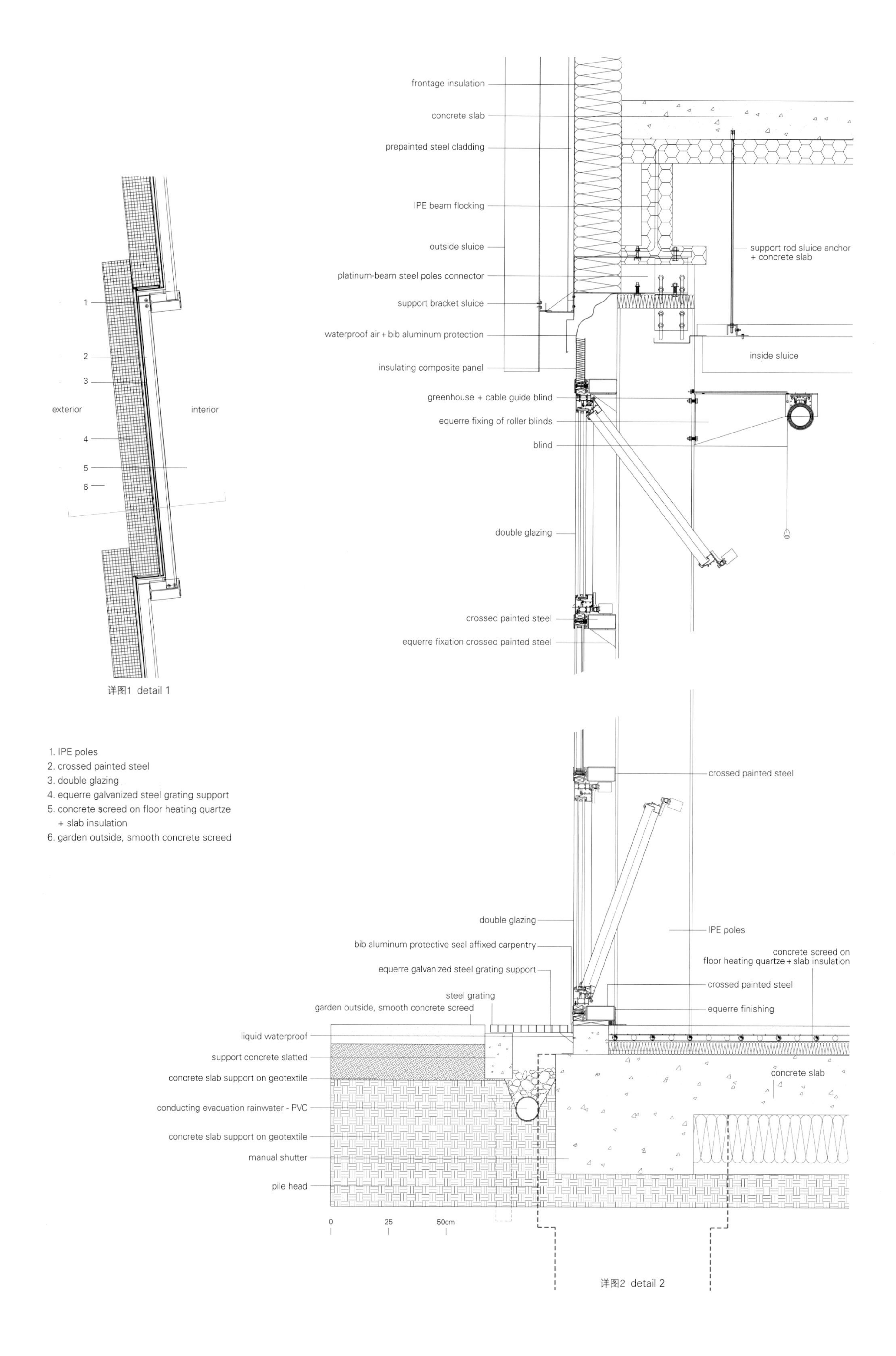

详图1 detail 1

1. IPE poles
2. crossed painted steel
3. double glazing
4. equerre galvanized steel grating support
5. concrete screed on floor heating quartze + slab insulation
6. garden outside, smooth concrete screed

详图2 detail 2

make up the edifice is scalable and identifiable by the color-material of the metal of the celestial body that is associated with it. So, the world of "creating" is anthracite in reference to Saturn and lead. The world of "understanding" recalls the moon and silver. The world of "imagining" is represented by Jupiter and bronze. The sun and gold are to be found in the world "from one world to another". Finally, the world "manufacture" is red copper, paying homage to Mars.
Connotations of the imaginary world of childhood are also seen in the design and the furniture which is geometrical, fun and functional, responding to the different uses that the multimedia library offers. Wooden frames around individual desks isolate the reader in an atmosphere that lends itself to concentration that at the same time, opens a window to the outside. Massive naturally lit tables encourage exchanges during group work. Indoor and outdoor seats are in the shape of extra-large objects, for example a safety pin or a peg, providing seating that is fun, friendly and modulable. The architect has freed up the space under the ceiling in the entrance hall by placing the large East staircase in levitation, as soon as one enters the building, a space that each individual can colonise and own. Additionally, the spaces are extended outside with terraces or gardens, for indoor/outdoor living, to make the most of natural light everywhere and to open up or close off the spaces depending on the heat levels and mount of light necessary or desired. Based on the observation that the city dweller is increasingly cut off from nature and finds in the elements of life and nature a rich source of dreams and of pleasure, Loci Anima has, here again, created this building as a space for the living. Sun, air, plants, water and metal are all materials used in the construction of the project. Over and above the real and intrinsic performances of the building on environmental aspects (low energy consumption buildings), this way of seeing, which is specific to Loci Anima, instils a symbolic and sensitive dimension to the project.

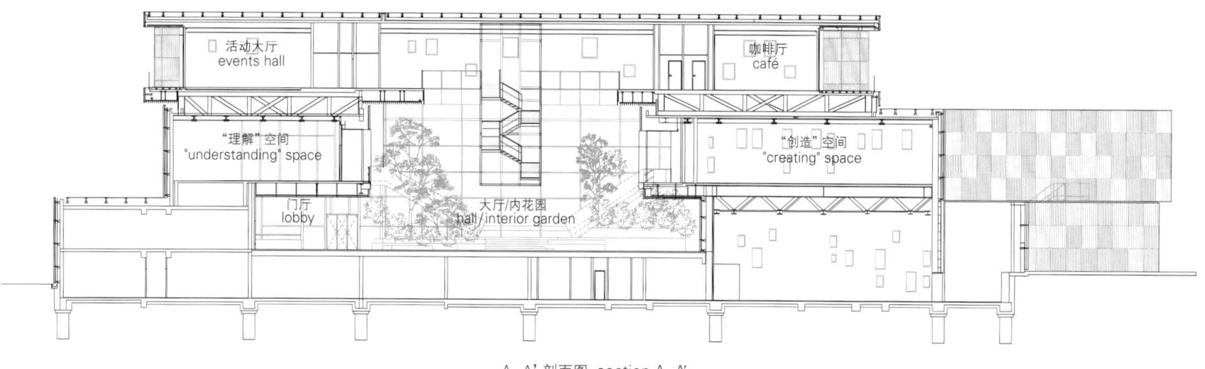

A–A' 剖面图 section A-A'

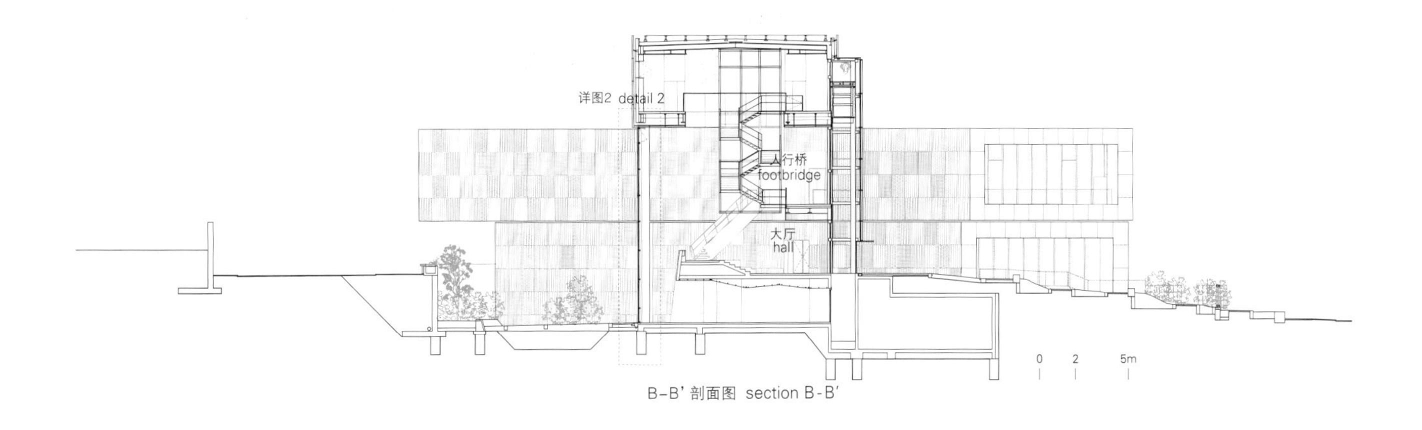

B-B' 剖面图 section B-B'

项目名称：Alpha Multimedia Library / 地点：Angoulême, (16) France / 建筑师：Françoise Raynaud _ Loci Anima / 项目经理：Xavier Maunoury / 项目总监：Jonathan Thornhill / 施工经理：Marine Bichot / 结构与暖通空调工程师：Grontmij / 立面工程师：Van Santen & Associés / 环境工程师：Alto / 景观设计师：Exit / 声效工程师：Avel / 安全工程师：Casso & associés / 工料测量师与现场指导：Grontmij / 建筑管理人员：Alpes Controle / 现场协调：Ouest Coordination / 现场保卫：Socotec / 环境顾问：Addenda / 客户：Community of Greater Angoulême Agglomeration / 用途：Media Library, indoor/outdoor garden, amphitheatre, café and restaurant / 建筑面积：5,241 sqm / 造价：EUR 13.8 M / 竣工时间：2015.12 / 摄影师：©Philippe Le Roy (courtesy of the architect)

作家剧院

Studio Gang Architects

作家剧院是一家深受欢迎的戏剧公司，位于美国伊利诺伊州芝加哥郊区的小镇格伦科。该公司于 1992 年在一家书店的库房创立，创立之初就尊崇亲切宜人的审美理念。2003 年，作家剧院在格伦科女子图书馆俱乐部创建了一个更大的场地，但是这个拥有 108 个座位的新剧场不久就遭遇了空间有限的挑战。尽管拥有一批热情的忠实观众，但是公司仍然面临惨淡的经济前景。他们每晚的观众都接近爆满，但是座位太少，加上生产成本日益增加，考虑到自身发展，他们需要一个更大、更加灵活的空间。

就在此时，格伦科——一个位于芝加哥以北约 32.2km、人口近 9000 人的郊区小镇——开始了一个宏大的总体规划，力图在市中心整合更多的文化和商业空间。同时，现在这个破旧的女子图书馆俱乐部也亟需修缮。作家剧院与女子图书馆俱乐部和格伦科镇委会合作，抓住机会，在图书馆俱乐部的原址上建造了一个量身打造的戏剧中心，以促进该地区市中心的发展。该公司每个季度的观众都能达到 35000 人，其一贯的高品质以及亲民的艺术风格获得了广泛赞誉。

对于这个永久的新家，作家剧院既想保留原来的亲切宜人的审美理念，又希望能够适应观众群体不断增长的需求，为来自全球各地的艺术家改善剧院设施，并与公众创建新型关系。由此所形成的设计，以其通透性和灵活性为格伦科市中心的日常生活注入活力，营造了一个开放、友好的空间。在这里，明显可以看出剧院通过经验分享将所有人团结起来的潜力。

剧院的入口大厅位于一层，为公众聚集之所，同时也是演出、排练、观众外联项目的非正式场地以及大型演出活动的正式场地。共有两个表演场地：一个是有 250 个席位的主舞台，另一个黑盒子表演场地有 99 个席位，还有排练室及其他公共设施，都朝向这个中央空间开放。二层的走廊通道采用空腹木桁架结构，浅色的木格架紧挂在主体结构上，可欣赏市中心、湖泊以及附近小树林的景色。当天气温暖的时候，剧院这个中心枢纽的影响力可辐射到邻近的公园和市中心，继而将其

交流性与活力传递到其他社区。在晚上，剧院就像一个灯笼，从里向外，熠熠生辉，吸引人们来到这个重要的文化活动场所和格伦科市中心参加各种活动。表演场地的设计意在最大程度提升演员与观众之间的亲密感，同时加强剧院自身特色：给观众带来身临其境的体验。在较大的表演场地中，看台的座椅没有紧靠墙体，而是留出空间，这样，演员可以采用令人兴奋的入场方式，也可以带来其他富有创意的舞台效果，而剧场地面与舞台的无缝过渡进一步加强了上述效果。为了向作家剧院公司的历史致敬，用从女子图书馆俱乐部拆下的砖建了一个样式精巧的后台隔声屏，用来扩散和反射声音，营造亲切的声环境。而另一个较小的黑盒子表演场地可以完全按照不同的表演和活动要求灵活布置。屋顶的凉亭和绿化屋顶也提供了额外的活动空间。

作家剧院不仅服务于格伦科当地社区，还吸引了芝加哥市内以及更远地区的观众。随着这个新的文化场所的开放，格伦科每年将吸引游客近 45000 人次，人们在此共同体验作家剧院的演出、社区举办的活动、工作坊和聚会，将表演艺术融入到日常生活中。

Writers Theater

Writers Theater is a popular theater company in the Chicago suburb of Glencoe, Illinois. Founded in the back room of a bookstore in 1992, Writers has embraced intimacy as its hallmark aesthetic since the very beginning. In 2003, Writers established a larger space at the Woman's Library Club of Glencoe, but, at 108 seats, this new venue soon imposed challenges of its own. The company faced bleak economic prospects despite an enthusiastic and committed audience. They were playing close to capacity night after night, but with very few seats to sell and production costs steadily rising, they were in need of a larger, more flexible space to allow for their growth.

Meanwhile, Glencoe – a suburb 20 miles north of Chicago with a population of approximately 9,000 residents – embarked on an ambitious master plan to integrate more cultural and commercial spaces in the downtown area. At the same time, the existing but deteriorated Woman's Library Club building was in serious need of repair. Partnering with the Woman's Library Club and the Village of Glencoe, Writers seized the opportunity to build a custom theater center and catalyst for downtown development on the Library Club site. The company, which plays to an audience of 35,000 patrons each season, has garnered critical praise for the consistent high quality and intimacy of its artistry.

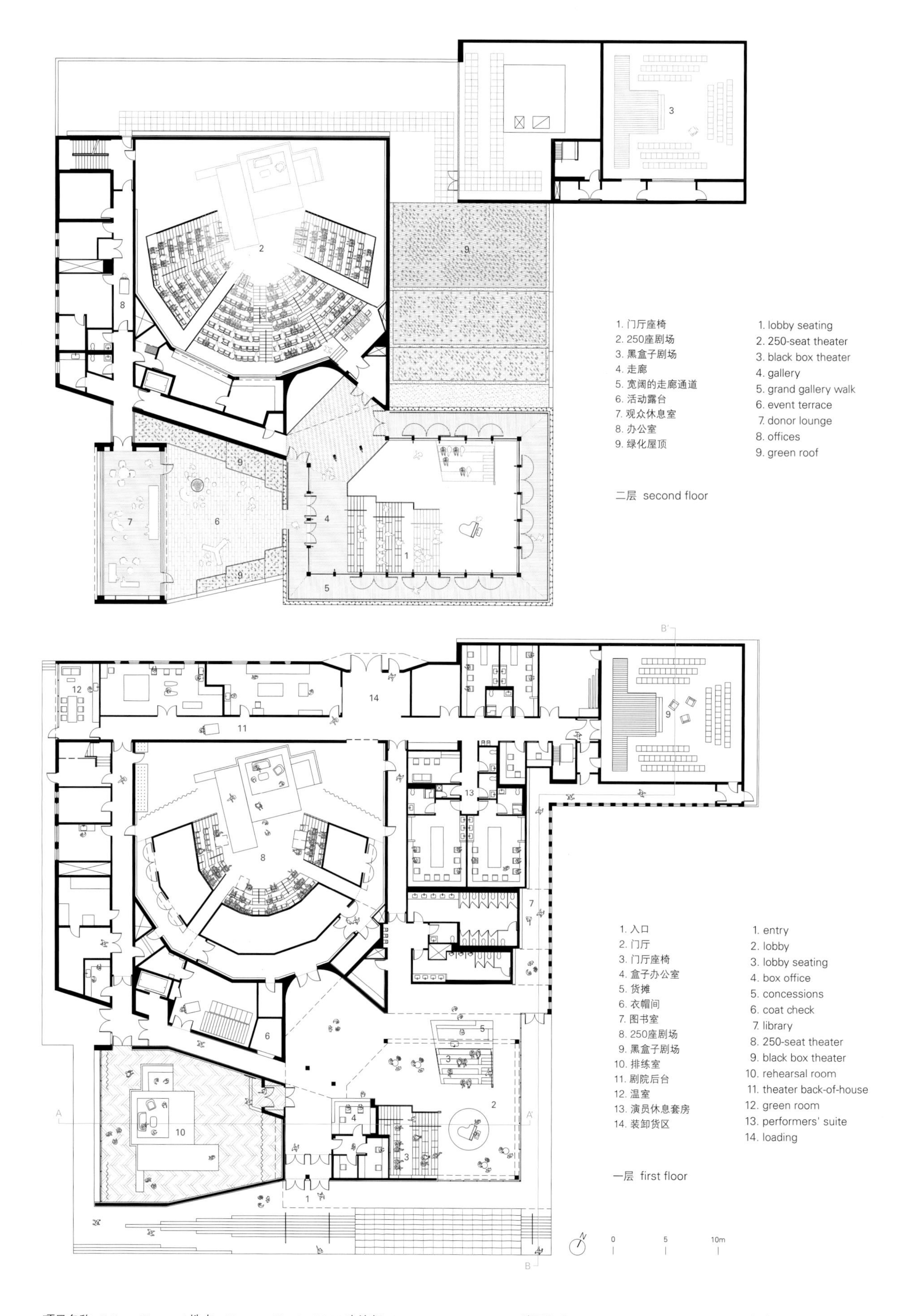

项目名称：Writers Theater / 地点：Glencoe, Illinois, USA / 建筑师：Studio Gang Architects / 总承包商：W.E. O'Neil Construction / 业主法律顾问：AMS Planning & Research Corporation and VMS LLC / 剧场顾问：Auerbach Pollock Friedlander / 景观设计师：Coen + Partners / 机电、防火工程师：dbHMS / 结构工程师：Halvorson and Partners / 照明设计顾问：Lightswitch Architectural / 宽阔走廊通道的工程专家：Peter Heppel Associates / 土木工程师：SPACECO, Inc. / 平面设计师：Thirst / 声效顾问：Threshold Acoustics / 造价顾问：Venue / LEED/可持续性设计顾问：WMA Sustainability Solutions Group / 客户：Writers Theater / 用途：Cultural, Office / 用地面积：4,552m² / 总建筑面积：3,345m² / 造价：$28 million / 项目开始时间：2012 / 竣工时间：2016.2 / 摄影师：©Steve Hall-Hedrich Blessing (courtesy of the architect)

WOOLF: A PARODY APR 27 - JUL 17
A STREETCAR

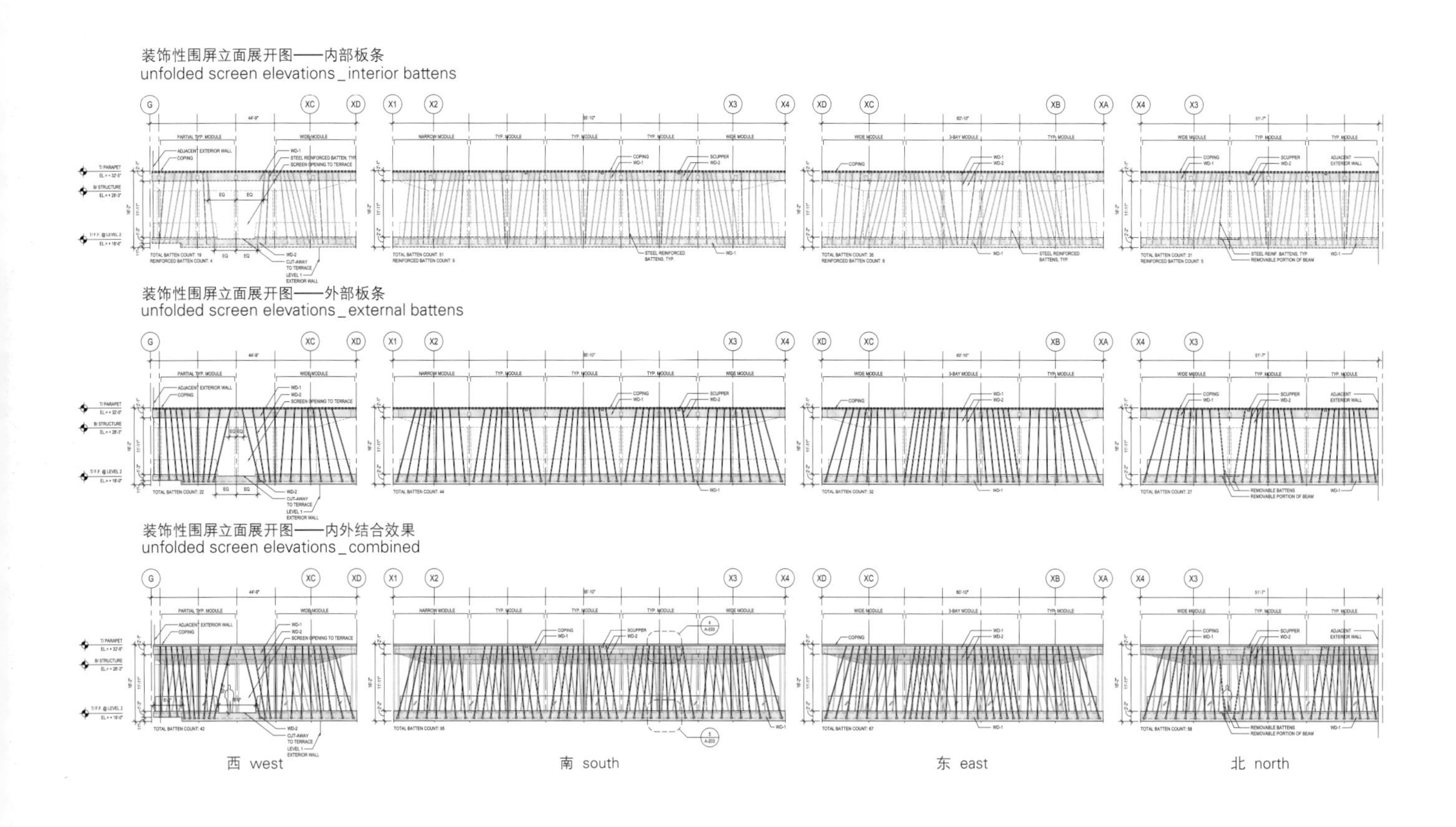

装饰性围屏立面展开图——内部板条
unfolded screen elevations_interior battens

装饰性围屏立面展开图——外部板条
unfolded screen elevations_external battens

装饰性围屏立面展开图——内外结合效果
unfolded screen elevations_combined

西 west　　南 south　　东 east　　北 north

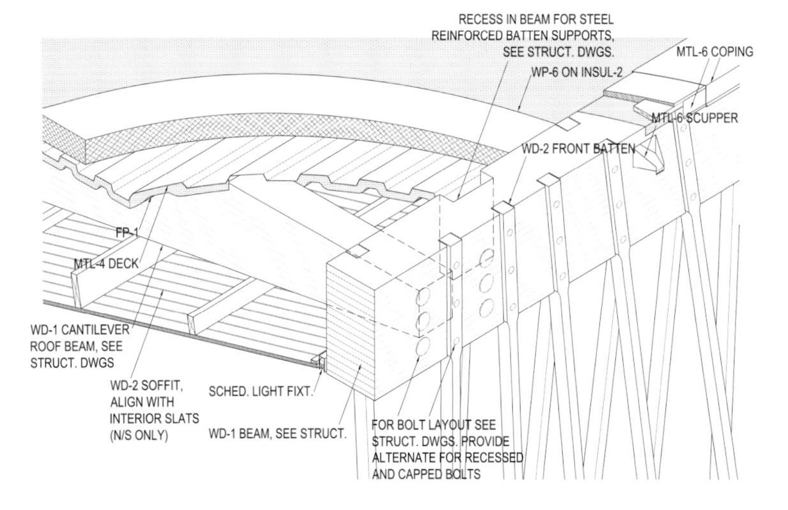

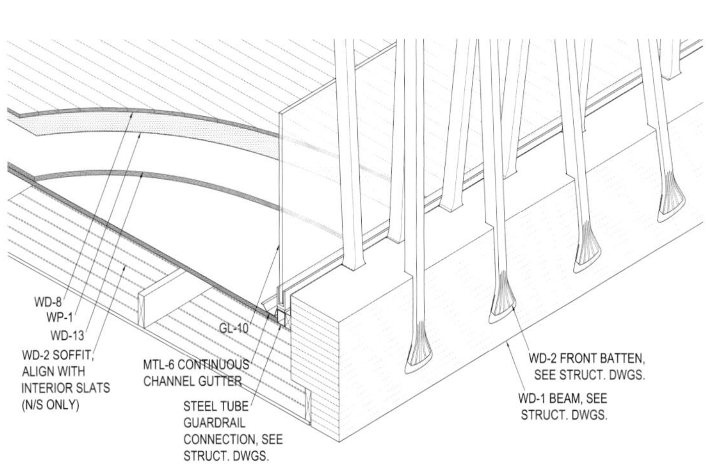

装饰性围屏详细轴测图
screen detail axonometric

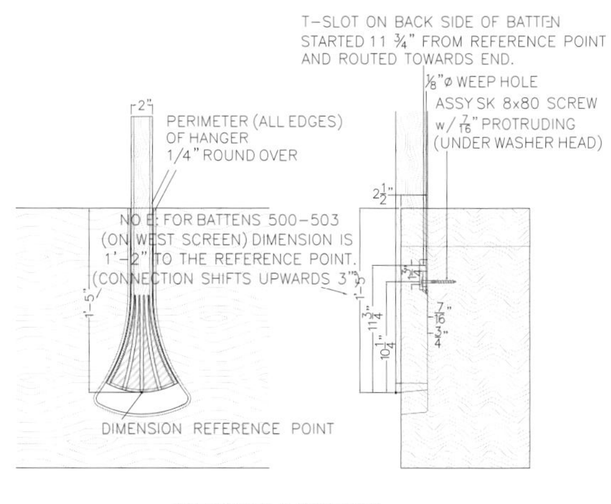

所有外部挂钩的底部连接处
bottom connection for all exterior hangers

WRITERS
THEATRE

With their new permanent home, Writers wanted to ensure these hallmark traits were maintained while also accommodating a growing audience base, improving their facilities for their global community of artists, and creating new relationships with the public. The resulting design, with its transparency and flexibility, is intended to energize daily life in downtown Glencoe, creating an open, welcoming space where the potential of theater to unite people across boundaries through shared experience is rendered visible.

The theater is anchored by a public gathering space that serves as a lobby as well as an informal space for performance, rehearsals, and audience outreach programs, and a formal space for events. Two performance venues, a 250-seat main stage and a 99-seat black box venue, as well as rehearsal rooms and other public amenities, open to this central space. A second-floor gallery walk, providing views toward the downtown, lake, and nearby grove, is structured by timber Vierendeel trusses and a lighter wood lattice hung in tension from the primary structure. In warm weather this central hub opens to the adjacent public park and downtown, allowing the energy and interaction generated within the theater to extend outward into the community beyond. At night, it glows from within like a lantern, drawing interest and activity to this important cultural anchor and downtown Glencoe.

The design of the performance spaces is intended to maximize the sense of intimacy between actors and audience and enhance the immersive experience for which Writers is known. In the larger venue, tribune seating is liberated from the walls, inviting exciting actor entrances and other innovative staging opportunities, further enhanced by a seamless transition from theater floor to stage. In a nod to the history of the company, bricks reclaimed from the Woman's Library Club building form an elaborately patterned back-of-house acoustic screen that diffuses and reflects sound for an intimate aural environment. The smaller black box venue can be infinitely customized for performances and events. A rooftop pavilion and green roof offer additional event space.

Writers Theater not only serves its immediate Glencoe community but also attracts audiences from the Chicago metropolitan region and beyond. With the opening of this new cultural facility, nearly 45,000 additional people could be drawn to Glencoe each year to share in the experience of the company's performances, community events, workshops, and gatherings, infusing the art of performance into their everyday lives.

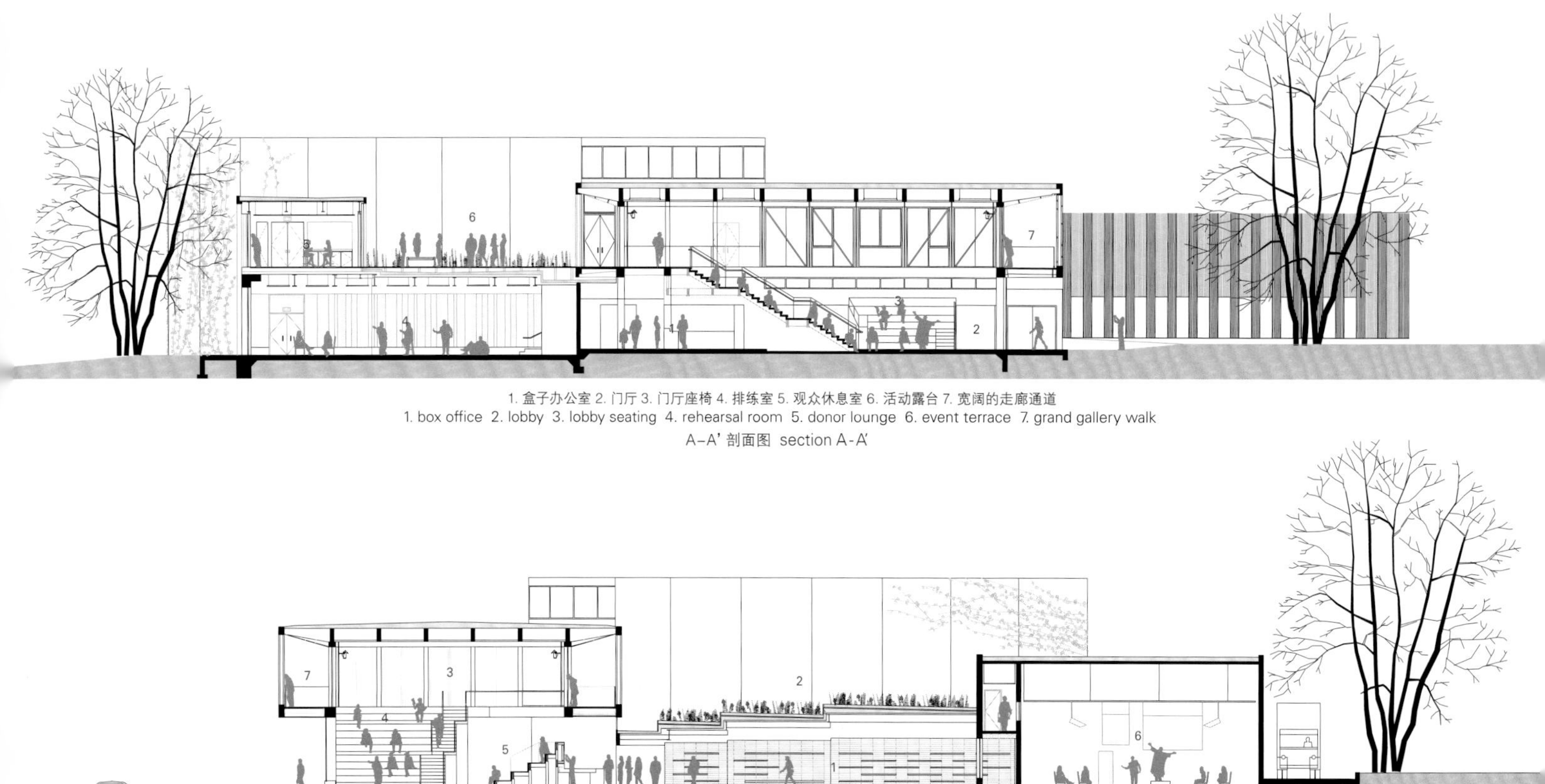

1. 盒子办公室 2. 门厅 3. 门厅座椅 4. 排练室 5. 观众休息室 6. 活动露台 7. 宽阔的走廊通道
1. box office 2. lobby 3. lobby seating 4. rehearsal room 5. donor lounge 6. event terrace 7. grand gallery walk
A–A' 剖面图 section A-A'

1. 图书室 2. 绿化屋顶 3. 门厅 4. 门厅座椅 5. 货摊 6. 黑盒子剧场 7. 宽阔的走廊通道
1. library 2. green roof 3. lobby 4. lobby seating 5. concessions 6. black box theater 7. grand gallery walk
B–B' 剖面图 section B-B'

WT CONCESSIONS CENTER

EXIT

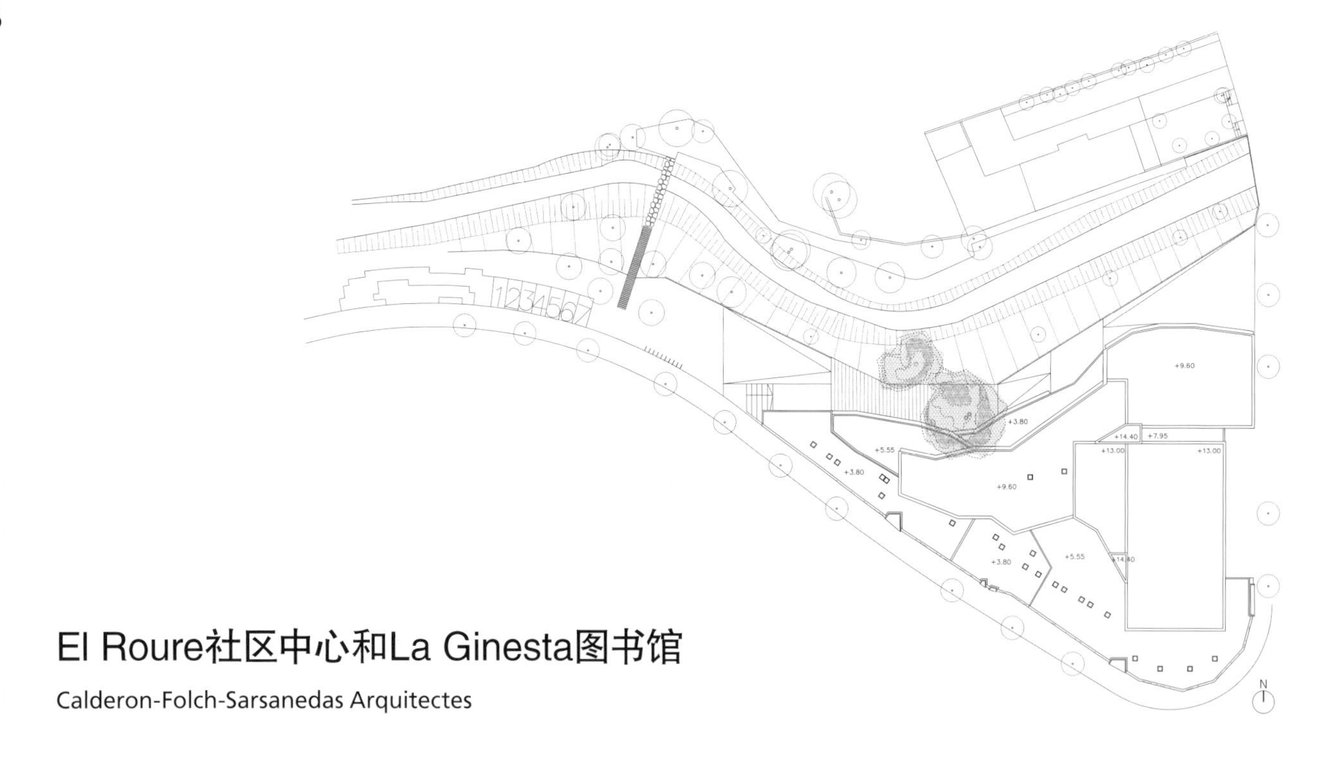

El Roure社区中心和La Ginesta图书馆

Calderon-Folch-Sarsanedas Arquitectes

贝格斯是位于格拉夫自然公园中的一个自治市，需要修建一座图书馆、一个社区中心和一个多功能剧院。这是一个由当地几家公共机构和三个行政部门共同参与的项目。为了实施这一项目，项目地点最终选定了贝格斯溪边的一个三角形地块。

该新建筑的设计基于如下两个主要理念：一是创造一个聚集人脉的"内部广场"，二是与环境融为一体，恢复贝格斯溪的活力。

"内部广场"：本项目是集三种服务设施于一体的单体建筑，创造了一个人们在此碰面的社区空间。这一空间将增进市民互动、增强文化整合，并提高在建造和管理方面的可持续性。这样的建筑应该多元化、多中心，以支持各种各样的用户和活动需求，但还必须能够加强、协调不同机构和使用者之间的关系。该项目在一开始就确定了其重要的核心理念：建造一个城市广场，能够吸引和连接这个功能项目中所定义的所有区域，也就是一个让所有使用者都各得其所的地方。

与环境融为一体，恢复贝格斯溪的活力：这个三角形地块位于一个长满松树的小山丘的山脚，三角形的两条短边由两条街道围住，而长边毗邻丰达溪（在加泰罗尼亚水事厅起草的防汛书上如此明确标注）。

建筑设计接纳了这个三角形地块自身的局限性，因势利导，对整个地块进行充分利用，呈横向建筑外形，与溪流景观融为一体，而门厅也位于这一侧。

设计想要复原溪流景观和当地人对池塘的共同记忆（一个夏季休闲娱乐的好去处）。为此，建筑主立面向北，吸引从拉尔路上经过的社区居民来欣赏这处被遗忘的风景，复兴生态系统，并促进与社区中心使用者之间新关系的建立。因此建筑的设计既像溪流，又像蜿蜒的河流和池塘，既有流动感，又映衬着周边环境。建筑主立面的几何外形浑然天成，与蜿蜒的人行步道相互映衬，既修建了一条新的通道，又将那棵橡树环抱其中（社区中心因它而得名）；建筑材料既有玻璃（建筑物的每片玻璃都能够倒映风景），也有来自生物圈的材料（尊重环境的自然原始风貌）。

在内部，项目的每个部分都有其自然位置，几乎完全遵循水流的自然运行方式：河岸是由水流形成的侵蚀和沉淀形成的。因此，建筑内部是以纵向层次组织的，描摹了溪流的线条，形成了有点加宽的空间，呈一种渐进的形态。靠近溪流一侧立面显得更加生动、流畅、清澈，水汪汪的；而面对山的那一侧立面则更加坚实、不透光、私密，使用了石材。朝向开阔风景那面的主要是图书馆，还有 EspaiNou、Punt Jove 和酒吧，

详图1 detail 1

详图2 detail 2

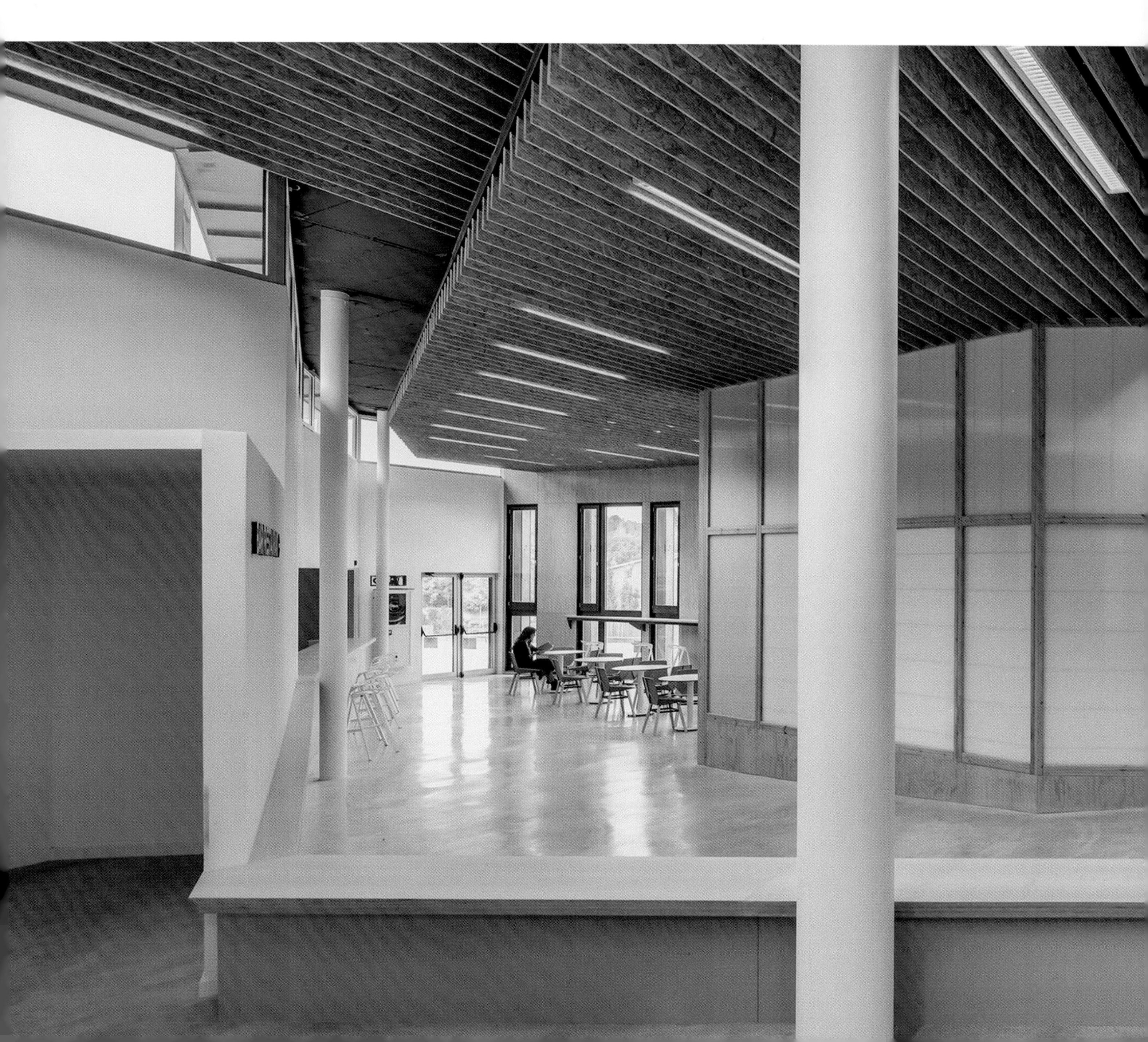

在这些地方人们可以欣赏到秀丽的景色。剧院、排练室以及各种较为封闭的服务设施都设在最后面或者距离溪流最远的位置。"内部广场"将所有这些空间连为一体，"内部广场"的形态、所使用的材质以及自然采光，都营造了一种自然的氛围，使人产生一种漫步溪流岸边的联想。

El Roure Community Center and La Ginesta Library

Begues, a municipality located in Garraf Natural Park, required a library, a community center and a multipurpose theater. To carry out the project, in which several local organisations and three administrations are participating, a triangular plot which goes along the edge of Begues Stream was arranged.
Two main ideas underpin the conception of the new facility: to generate a confluence "inner square" and to tune into the environment revitalising the Stream.
The "inner square": The project outlines a single building which gathers the three services, creating a community space, a place where people meet, which will enhance citizen interaction, cultural synergy and sustainability in its construction and management. The architecture that hosts a facility of this kind should be diverse and pluricentric in order to support a wide variety of users and foreseen activities, but it must also have the ability to strengthen and harmonise the relationship between organisations and users. The architectural project begins with the definition of a foundation core, an agora able to attract and articulate around all areas defined in the functional programme, a place where all users can identify themselves as belonging to it.
To tune into the environment revitalising the Stream: The triangular plot is located at the bottom of a hill covered with pine trees and bordered by two streets on its minor sides and

AULA POLIVALENT

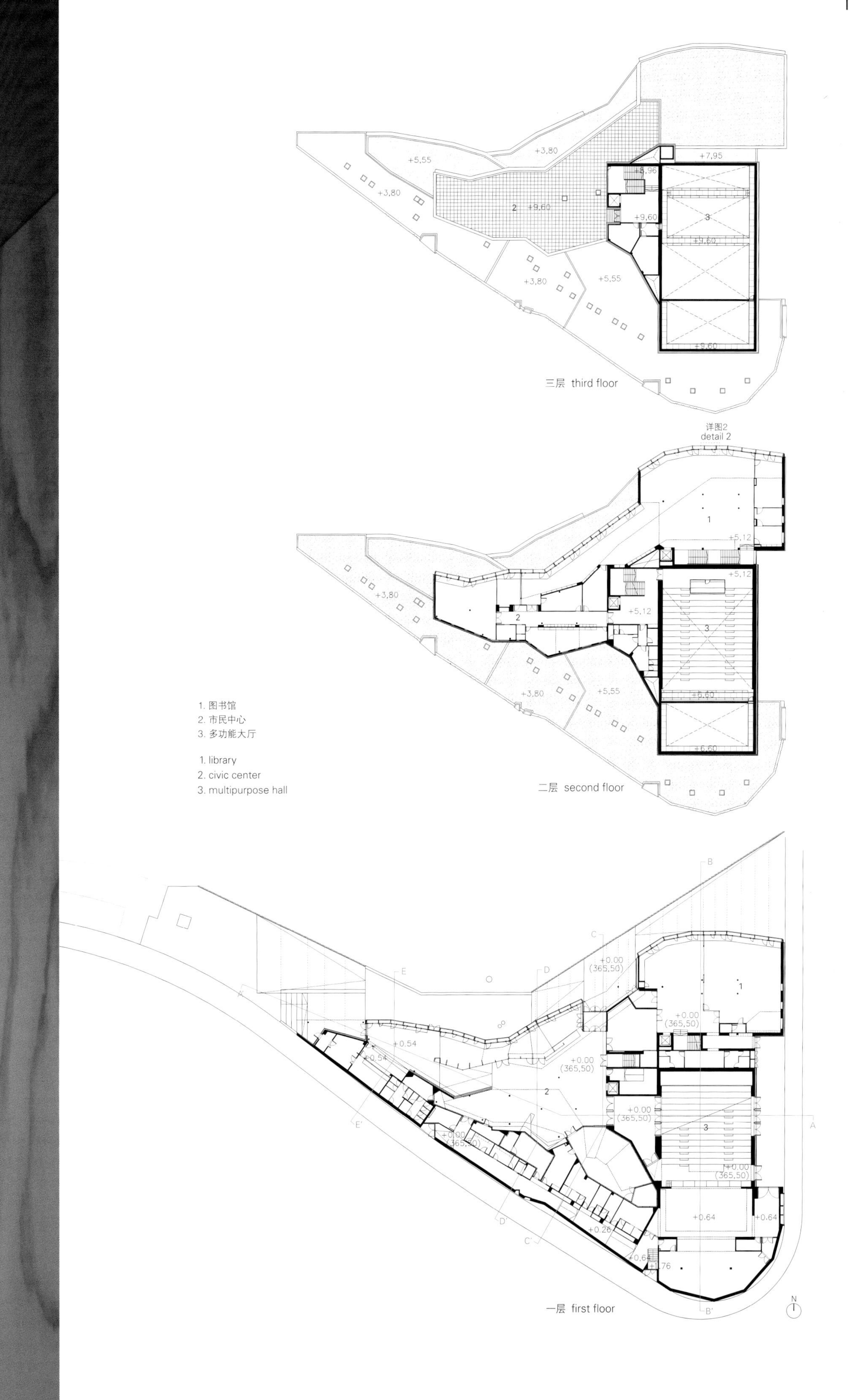

+5,55
+3,80
+7,95
+5,96
2 +9,60
+9,60
3
+3,80
+5,55
+9,60
三层 third floor
详图2
detail 2
1
+5,12
2
+5,12
3
+3,80
+5,55
+6,60
二层 second floor
1. 图书馆
2. 市民中心
3. 多功能大厅
1. library
2. civic center
3. multipurpose hall
+0.00
(365,50)
+0.54
2
3
+0.64
+0.26
+0.76
一层 first floor

by Fonda Stream (specifically on the flood limit drawn up by the Catalan Water Agency) on its major side.
The building accepts the limits of the plot as its own and occupies the entire place to achieve a horizontal construction integrated into the landscape of the stream and which can accommodate the lobby.
The design wants to recover the stream and the collective memory of a pool (popular place for summer recreation). For this reason, the main facade is oriented to the north, picking up the flow of neighbours coming from Ral path and reassessing this forgotten landscape, revitalising the ecosystem and promoting a new relationship with the users of the center.
The building therefore aims to be a stream, a meander and a pool, flowing and reflecting the environment. The profile of organic geometry of the facade reflects the reverberation of the winding meander, generating a new access path and embracing the oak that names the center; its materiality is sometimes mirrored (to reflect and multiply the landscape in

each piece of glass) and sometimes biospherical (to respect the naturalness of the environment).

Inside, each part of the programme finds its natural place, almost respecting the hydraulic logic by which erosion and sediment define the edge of the bank of the stream. Thus, the inside is organised in longitudinal layers tracing the line of the stream and resulting in more or less dilated spaces, which have a progressive materiality more dynamic, fluid, clear and watery near the facade of the stream and more solid, opaque, private or stony in front of the mountain. Mainly the library but also EspaiNou, Punt Jove and the bar are developed along this opening landscape facade which offers magnificent sightseeing options. The theater, the rehearsal boxes and the various more closed type services are in the last layer or farthest place of the stream. All these spaces are articulated through an agora whose morphology, materiality and natural lighting procure a natural atmosphere which recalls the one enjoyed touring the stream.

项目名称：El Roure Community Center and La Ginesta Library / 地点：Begues, Spain / 建筑师：Calderon-Folch-Sarsanedas Arquitectes
合作方：Ignasi Arbeloa_architect, Joan Vilanova_quantity surveyor, Marc Sanabra_structure, Anoche Iluminación Arquitectónica_lighting, María Retamero and Zoe Sarsanedas_graphic design, Eliseu Guillamon_landscape / 发起人：Town Hall of Begues / 施工方：IT2M /
室外面积：1,900 m² / 建筑面积：3,900 m² / 总建筑面积：3,893.15 m² / 设计时间：2009 / 施工时间：2014
摄影师：©Pol Viladoms (courtesy of the architect)

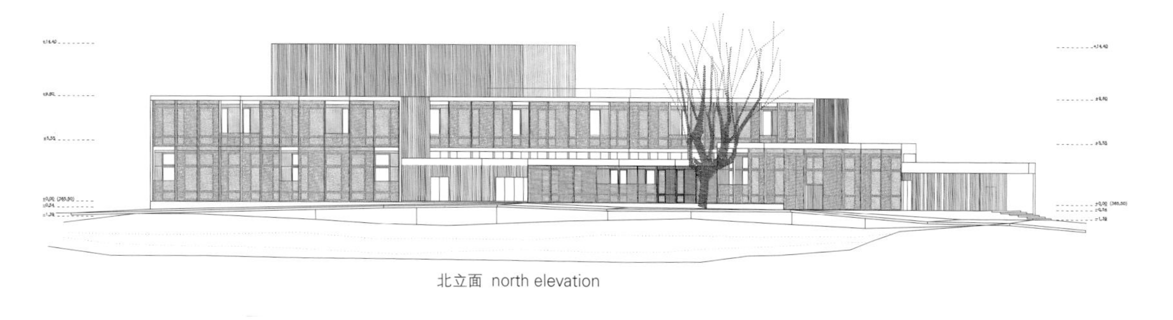

北立面 north elevation

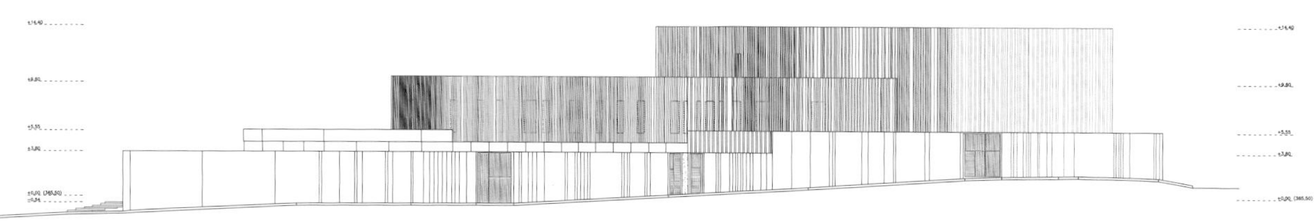

南立面 south elevation

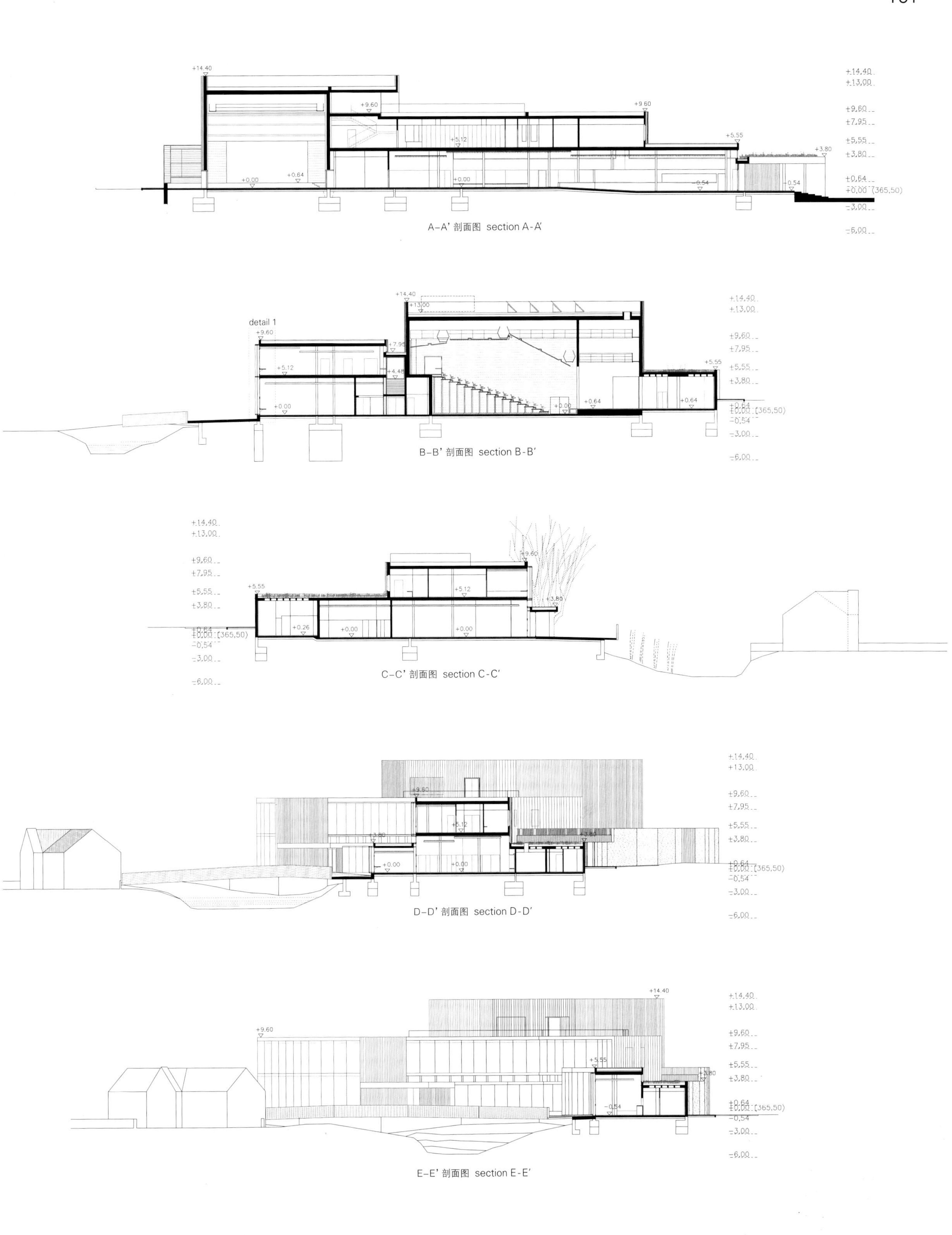

A-A' 剖面图 section A-A'

B-B' 剖面图 section B-B'

C-C' 剖面图 section C-C'

D-D' 剖面图 section D-D'

E-E' 剖面图 section E-E'

斯彻达尔文化中心

Reiulf Ramstad Architects

斯彻达尔文化中心打算取代一栋已有50年历史的老建筑。它因为面积太小，已不能满足城市快速发展的需要。斯彻达尔市是挪威的一个区域交通中心，位于瓦尔内斯的特隆赫姆机场的正北部，旁边就是特隆赫姆峡湾海岸，也是挪威海的一个入口。

项目的设计灵感基于本地区特有的历史和文化。与此同时，建筑设计也表现出了该建筑的现代功能和有节奏的时空变化。斯彻达尔文化中心被设计成一个令人欢欣鼓舞的地方，为游客提供各种体验和机遇，因而有助于丰富人们对当地文化和建筑的体验。

该文化中心项目无论在当地还是在整个地区都是一个重要的社区节点。对于所有对文化感兴趣的人来说（无论以何种方式），该中心都将是一个充满吸引力的地方。各式各样的人都可在此处挖掘或开发自己的能力与天赋。文化中心配有一座教堂，因此可以为居民举办各种仪式。此外，在酒店住宿的游客也将为该建筑和公园注入活力。斯彻达尔文化中心将成为一个广义的文化概念平台，包括艺术、舞蹈、音乐、电影和其他媒体等。中心也将成为一个鼓舞人心的地方，给游客带来不同体验，也为个人展示才能和发展提供机会。

文化中心共有三层，还有一个地下室，总建筑面积大约为11500m^2。斯彻达尔文化中心属于低层建筑，是传统的坡屋顶木结构，其设计重新诠释了当地特有的文化遗产，也与传统建筑规模相似。屋顶呈锯齿状，高度从14m到18m不等，教堂上方的屋顶呈尖塔形状，高高耸立。另外，该建筑有几处很有特色的悬臂式结构，从低层立面向外延伸出来，其中有的屋顶为人字屋顶。

Stjørdal Cultural Center

The Cultural Center is intended to replace an existing facility in Stjørdal that is 50 years old and considered too small for the fast-growing city. Stjørdal is a regional transportation center in Norway, located directly north of Trondheim Airport, Vaernes, and along the coast of the Trondheimsfjord, an inlet of the Norwegian Sea.

The project is anchored in and inspired by the history and culture of its location. At the same time, the architecture conveys its modern function and the pulse of the time and place. The Cultural Center of Stjørdal is designed to be an inspiring place that provides the visitors with experiences and opportunities,

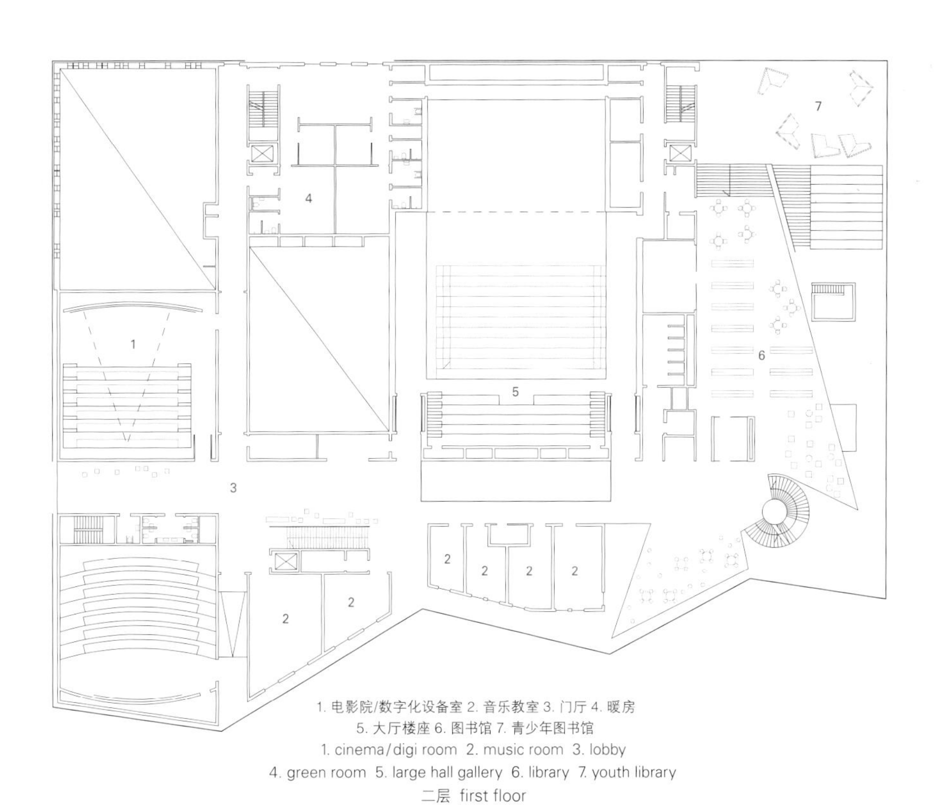

1. 电影院/数字化设备室 2. 音乐教室 3. 门厅 4. 暖房
5. 大厅楼座 6. 图书馆 7. 青少年图书馆
1. cinema/digi room 2. music room 3. lobby
4. green room 5. large hall gallery 6. library 7. youth library
二层 first floor

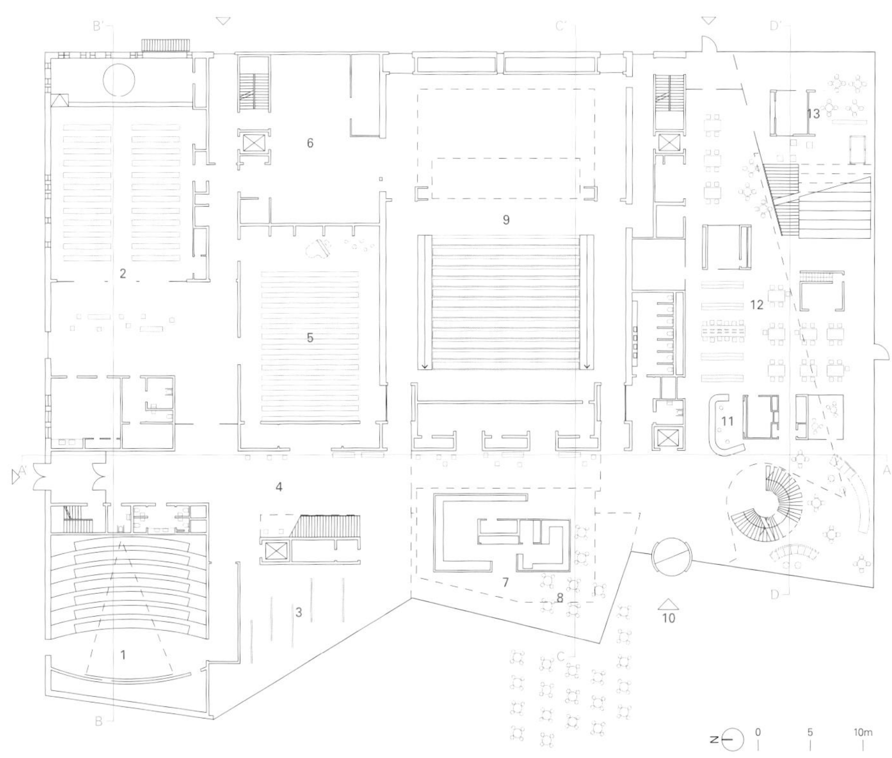

1. 电影院/数字化设备室 2. 教堂 3. 画廊 4. 门厅 5. 室内乐大厅 6. 储藏室 7. 商店
8. 咖啡厅 9. 大厅 10. 主入口 11. 接待处 12. 图书馆 13. 青少年活动室
1. cinema/digi room 2. church 3. gallery 4. lobby 5. chamber hall 6. storage 7. shop
8. café 9. large hall 10. main entrance 11. reception 12. library 13. youth house
一层 ground floor

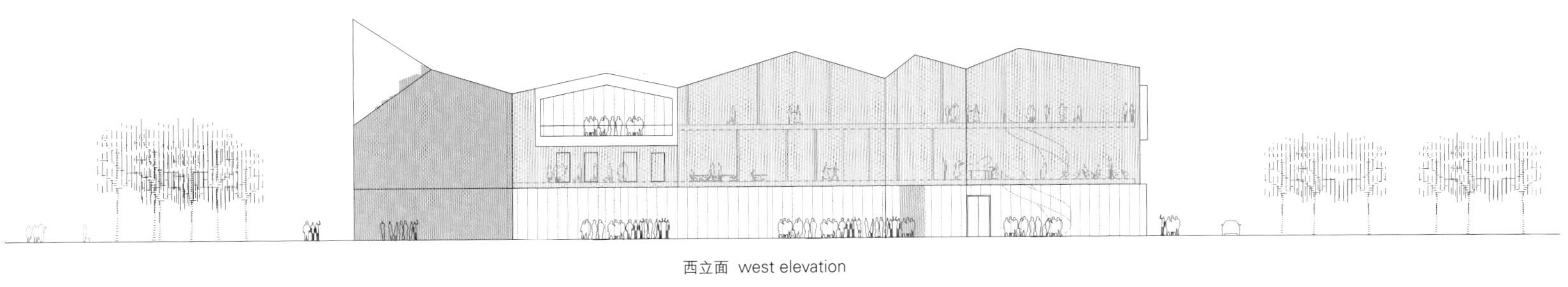

西立面 west elevation

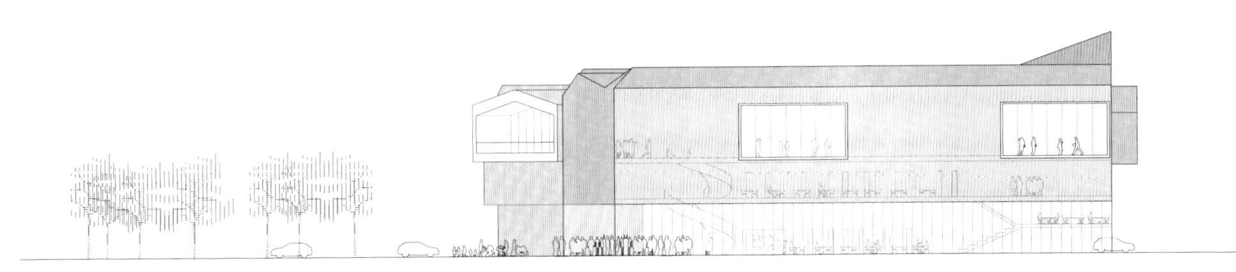

南立面 south elevation

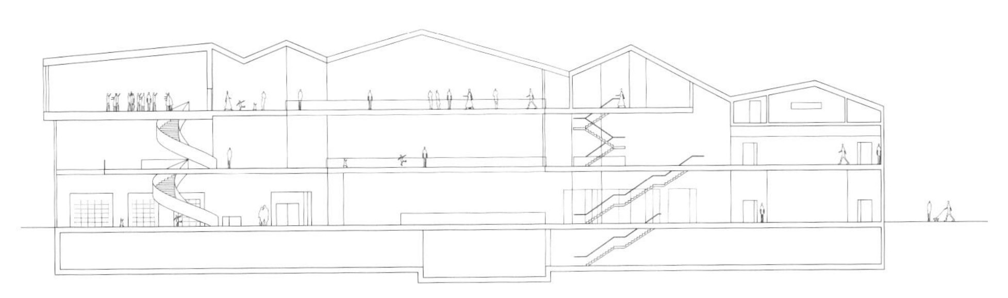

A-A' 剖面图 section A-A'

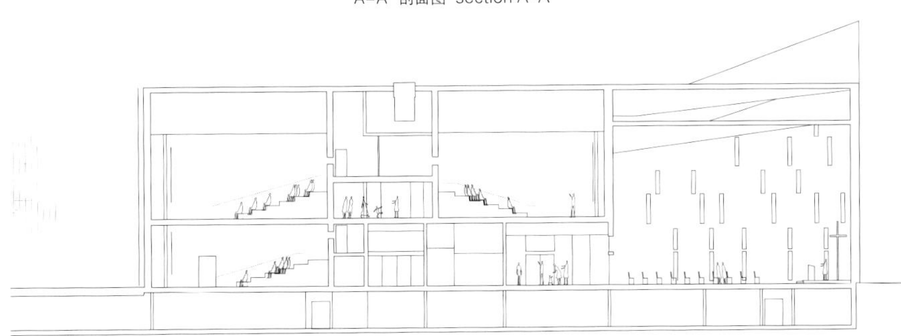

B-B' 剖面图 section B-B'

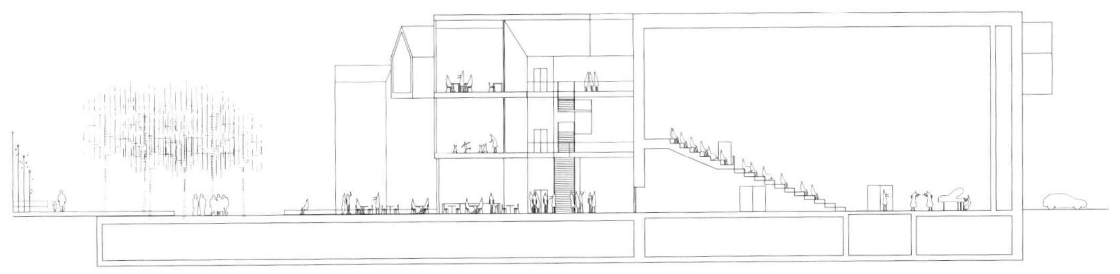

C-C' 剖面图 section C-C'

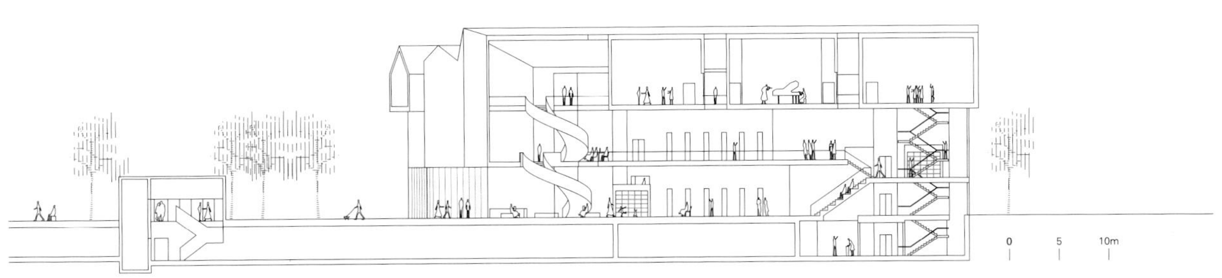

D-D' 剖面图 section D-D'

FREDERICK
FORSYTH

项目名称：Stjørdal Cultural Center / 地点：Stjørdal, Nord-Trøndelag, Norway / 建筑师：Reiulf Ramstad Arkitekter AS / 客户：Stjørdal Municipality and Stjørdal Kulturutvikling AS / 功能：New cultural center with concert halls, library, church, cinema, hotel, and the culture and music school of Stjørdal / 用地面积：6,500m² / 建筑面积：17,500m² / 总建筑面积：4,300m² / 造价：717 MNOK / 竞赛时间：2010 / 竣工时间：2015 / 摄影师：courtesy of the architect - p.176, p.178~179; ©Arne Wang (courtesy of the architect) - p.164~165, p.170~171; ©Søren Harder Nielsen (courtesy of the architect) - p.166, p.169, p.174, p.175, p.177; ©Wenzel Prokosch (courtesy of the architect) - p.168, p.172

thus contributing to a rich locational, cultural and architectural experience.
This project is an important node of the community, both locally and in the region. The center will become an inviting place for all people interested in culture in one way or another; a building where people of all kinds can explore and develop their abilities and talent. With its church, the Cultural Center will function as a worthy venue for all kinds of ceremonies for the inhabitants. In addition to this, the visitors at the hotel will contribute to vitalize the house and the park. The center will be a platform for a wide cultural concept; a wide range of art, such as dance, music, film and other media. The

building should become an inspiring place that gives the visitors experiences and opportunities for personal display and development.

The three-story center, which also has one basement level, provides a gross floor area of approximately 11,500 m².

The architecture of Stjørdal features low-rise buildings and traditional pitch-roofed timber structures, a heritage and scale that are being reinterpreted in the cultural center's design. The roof presents a saw-tooth form that varies in height from 14 to 18 m, the portion over the church rising in a spire. The structure also features several cantilevered sections that project out from the low-rise facade, some with gabled roofs.

Galtzaraborda停车库项目

VAUMM

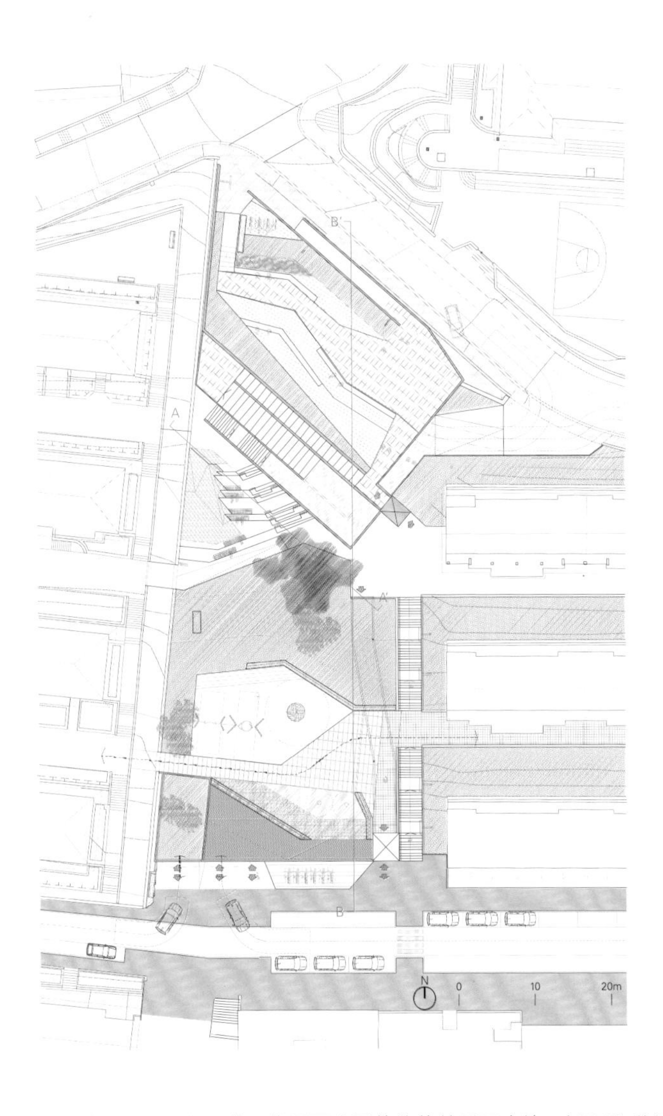

Galtzaraborda 居住区位于西班牙的埃伦特里亚小镇，建于 20 世纪 60 年代中叶。由于当地工业的快速发展，大批劳动力被吸引至此，住房需求大大提升，由此形成的城市化导致公共空间严重不足，且利用率低下，缺少无障碍环境。这些问题随着人口日益老龄化而不断加重。鉴于上述背景，这个停车场项目选在居住区的一个中心地块，这一地块在本市中尚未被开发利用。

停车场项目最初是为了解决停车问题，但其作用远不止于此，人们希望这个项目不仅能够解决停车问题，还能满足人们的出行需要，为人们提供公园和休闲娱乐之所。该项目共提供 111 个机动车停车位，分别布置在安装有全景电梯的两栋建筑中：一栋建筑一部电梯，还有两个大的屋顶平台。整个项目横向地将公民休闲娱乐设施和居民区连为一体，并确保居民区的最高点和最低点（火车站所在位置）垂直贯通。这个交通枢纽连接着埃伦特里亚镇和多诺斯蒂亚 – 圣塞巴斯蒂安地区。

该设施由两个半下沉的几何体组成，结合了场地的坡地地形，尽量减少了在审美方面和体积方面对该地块产生的影响。许多停车空间是开放的，因而设计了半透明的停车场立面，一改停车场给人留下的封闭沉闷的印象，人们也能直观地了解停车场的内部情况。坡道和楼梯使公园和相邻建筑之间的衔接连贯通畅，人们可以在停车场外面漫步。平台表面根据其用途采用了不同的处理方法，硬质路面与大型花园相互协调。绿化屋顶还原了建筑占地所缺失的绿化，将自然环境引入到这个高密度的都市环境中。

Galtzaraborda Parking

Galtzaraborda is a neighbourhood located in the town of Errenteria, Spain, that was built in the mid 1960s. Due to the great industrial development in the region, there was a huge demand for housing that was designed to absorb the labour migration wave, which generated an urbanism with serious deficiencies, inefficient amounts of public space, and a lack of accessibility. These problems have been accentuated by the progressive aging of the population. Intervening with this context, the project has been developed in a central plot, which was left undeveloped as a break in the fabric of the city.

The work, beyond resolving the parking, the function which motivates the project, is also conceived as an opportunity to generate an infrastructure that meets the circulation needs of the society, providing parks and leisure areas. The program of 111 automobile spaces is divided into two buildings with

two towers of panoramic elevators: one for each massing and two large squares as decks. Transversely, it connects civic amenities with the residential community and guarantees the vertical access from the highest point of the neighbourhood to the lowest level of it, where the train station is situated. This transportation hub links the town of Errenteria to Donostia-San Sebastián.

The facility is formed by two half-buried geometries, absorbing the slope of the site, generating the minimum aesthetic and volumetric impact. Many of the parking spaces are open, which allows the installation of a semi-transparent facade, avoiding a blind and inert front, and knowing intuitively the inner activity. A tour of the exterior perimeter can be made by qualities of the ramp and stairs, giving continuity to the transits through the park and the adjacent structures. The decks have different surface treatments according to their uses as hard pavement zones coexisting with large gardens. Green roofs return the original out areas occupied by the complex and introduce nature in a highly densified urban environment.

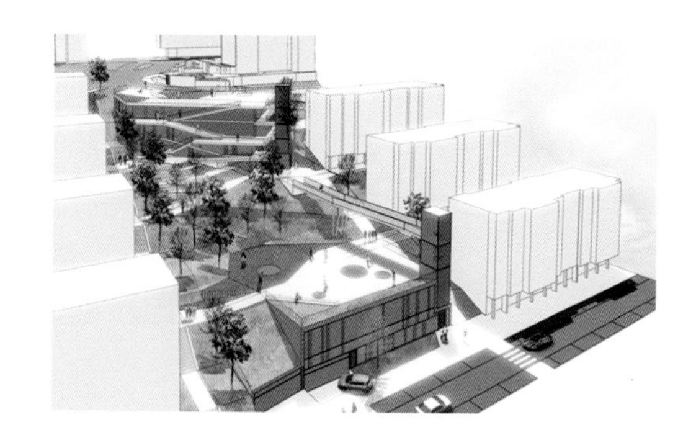

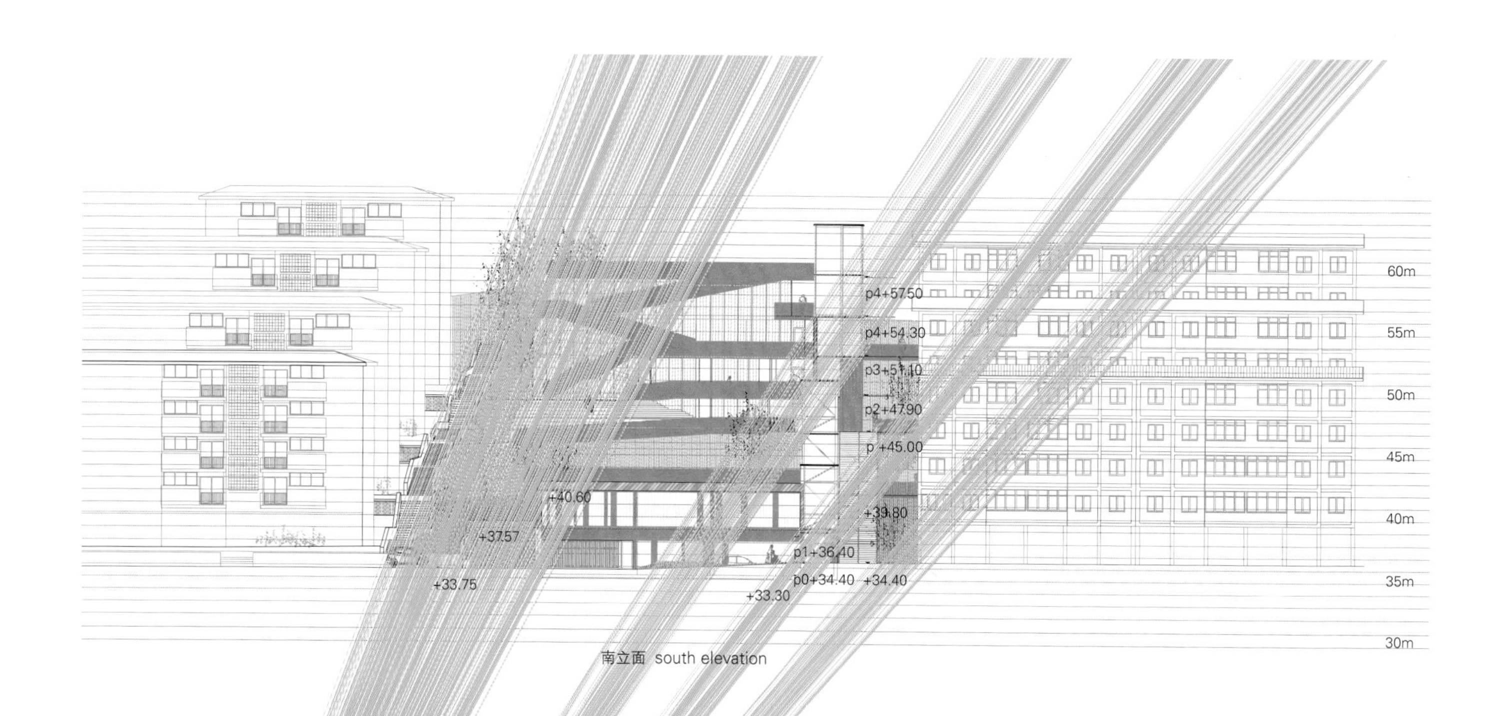

南立面 south elevation

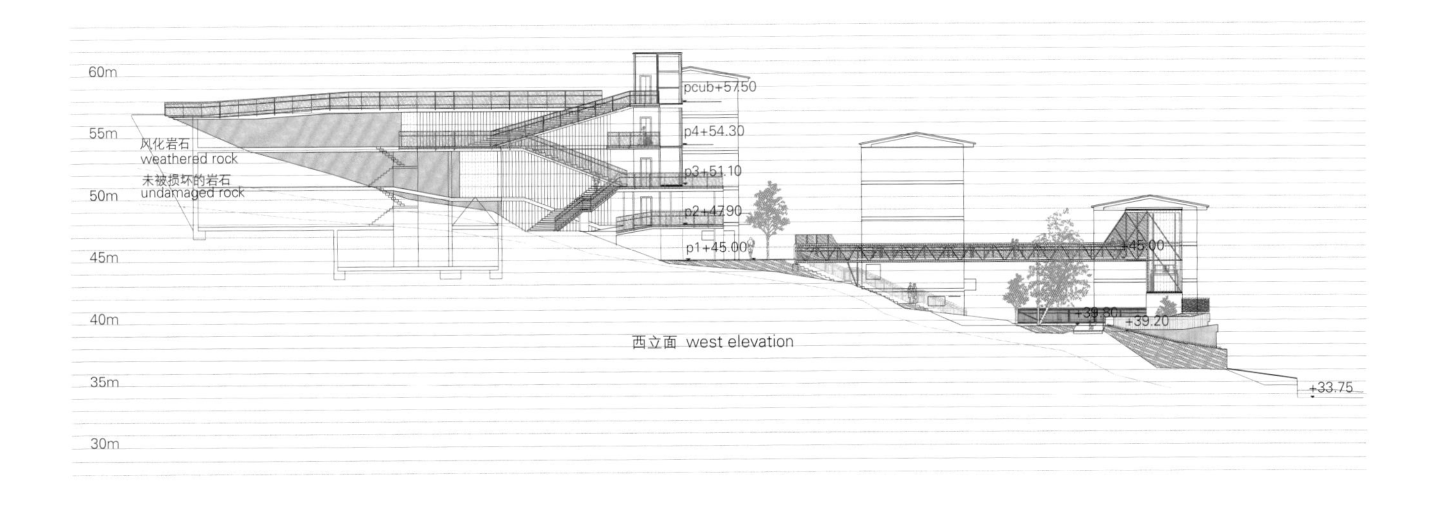

西立面 west elevation

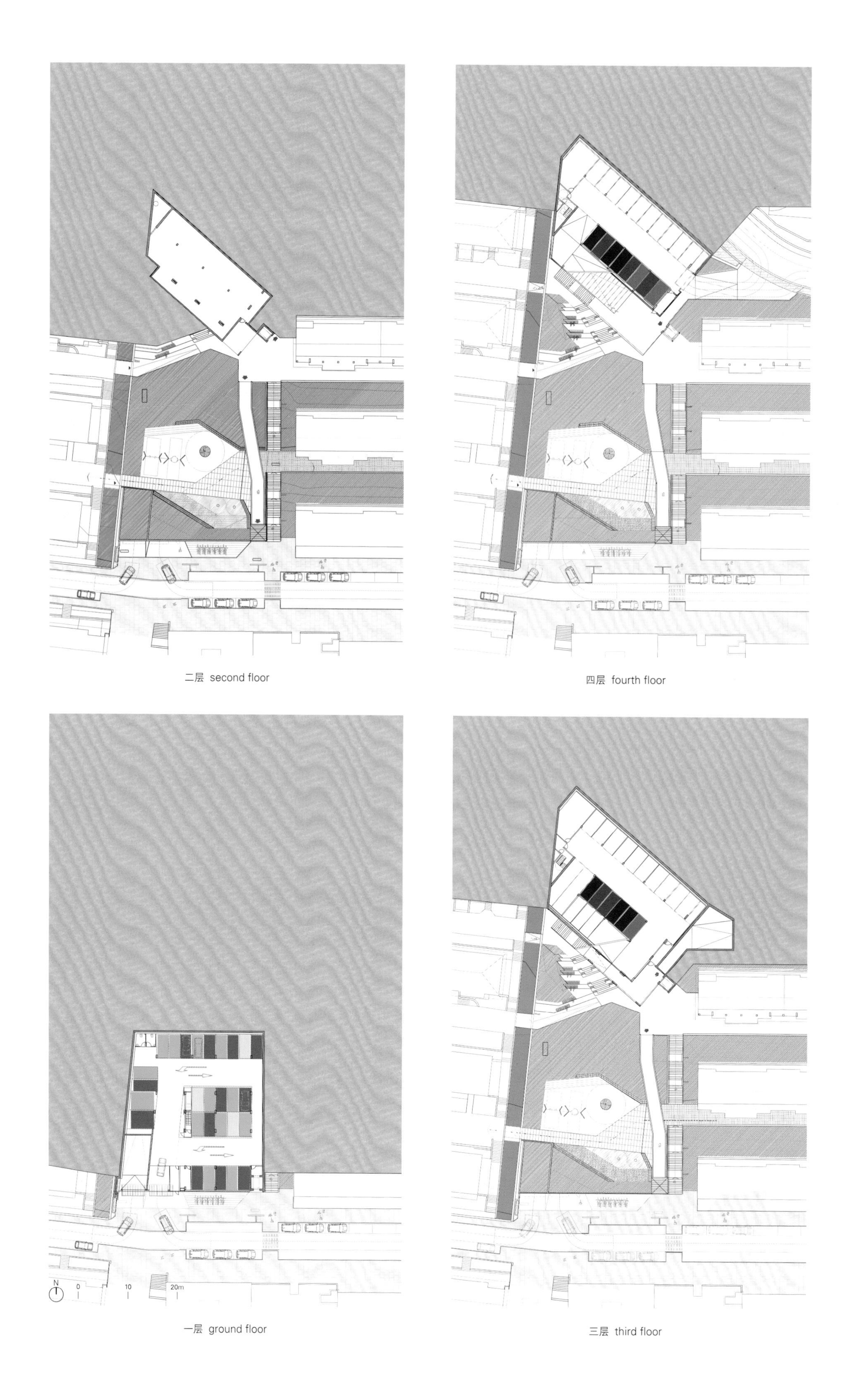

二层 second floor

四层 fourth floor

一层 ground floor

三层 third floor

项目名称：Galtzaraborda Parking / 地点：Errenteria, Gipuzkoa
建筑师：VAUMM architects
项目团队：Javier Ubillos Pernaut, Jon Muniategiandokoetxea Markiegi, Marta Alvarez Pastor, Iñigo García Odiaga, Tomas Valenciano Tamayo
施工方：UTE Olloquiegui-Tex (phase 1), UTE Ekolan- Alboka (phase 2)
技术架构师：Julen Rozas / 工程设计：Ingeniería de estructuras Lanchas
客户：Errenteria Town Hall
建筑面积：5,000m² / 造价：EUR 3,14 mill / 时间：2013
摄影师：©Aitor Ortiz (courtesy of the architect)

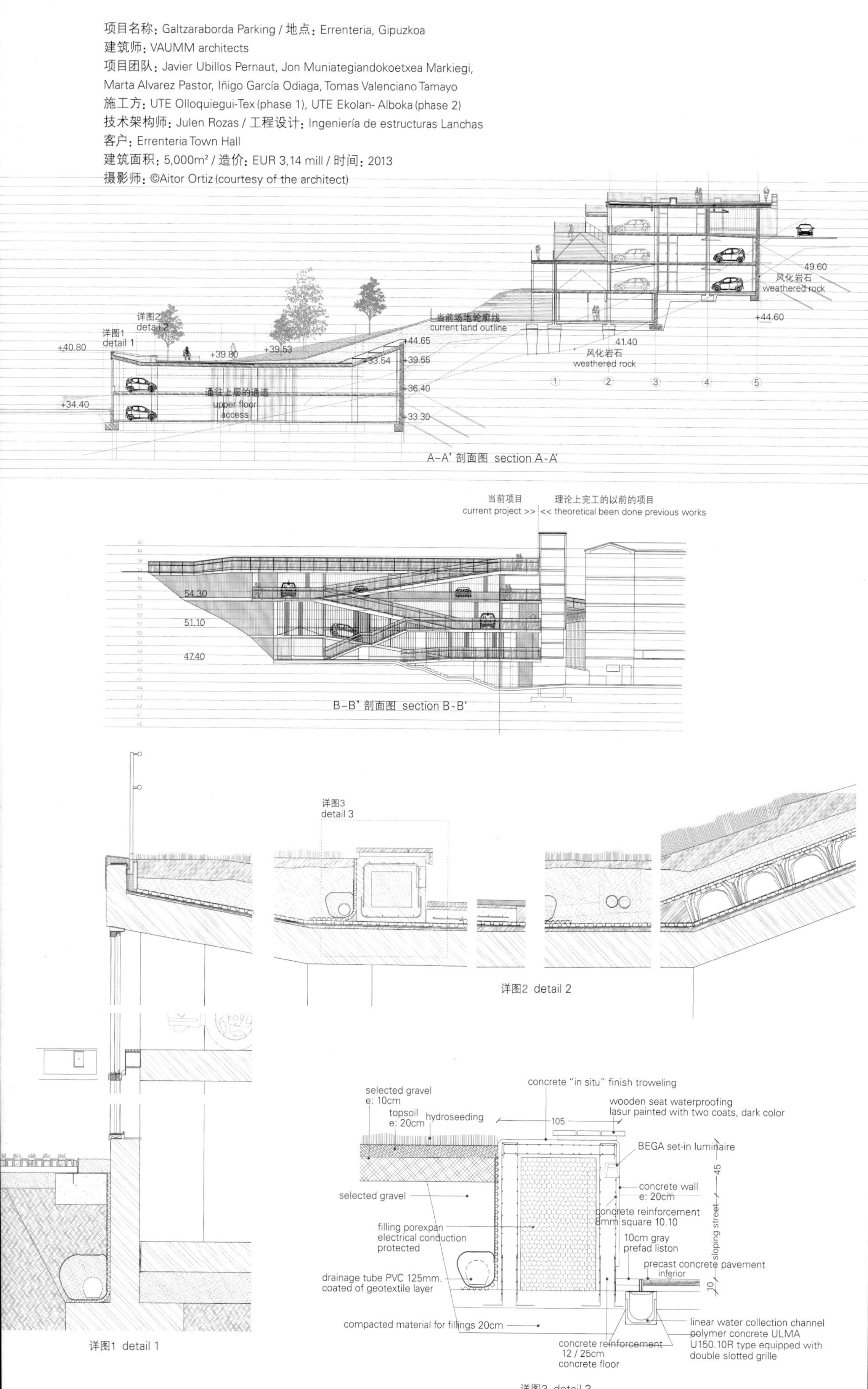

A-A' 剖面图 section A-A'

B-B' 剖面图 section B-B'

详图2 detail 2

详图1 detail 1

详图3 detail 3

Studio Gang Architects

Is a Chicago-based collective of architects, designers, and thinkers practicing internationally that was founded by MacArthur Fellow Jeanne Gang. She uses architecture as a medium of active response to contemporary issues and their impact on human experience. Each of her projects resonates with its specific site and culture while addressing larger global themes such as urbanization, climate, and sustainability. Her work has been exhibited at the Venice Architecture Biennale, the Museum of Modern Art, the National Building Museum, and the Art Institute of Chicago. She has received an Academy Award from the American Academy and Institute of Arts and Letters in 2006.

146

Calderon-Folch-Sarsanedas Arquitectes

Is an Architecture office in Barcelona founded in 2000. Founding partners, Pilar Calderon and Marc Folch received masters in Architecture at the ETSAB. Calderon completed her studies in the School of Architecture, Design and Urban Studies of the UC in Santiago of Chile, and Folch studied in the Lund Teknisker Högskolan of the Lund University in Sweden. Pol Sarsanedas studied at the Architecture Academy of Mendrisio and ETSAB. Joined Calderon-Folch in 2009, he became an associate architect at Calderon-Folch-Sarsanedas Arquitectes in 2011.

Schmidt Hammer Lassen Architects

Was founded in Aarhus, Denmark, in 1986 by architects Morten Schmidt, Bjarne Hammer and John F. Lassen. Is one of Scandinavia's most recognized and award-winning architectural practices with 30 years of experience. Now has Senior Partners Kim Holst Jensen and Kristian Lars Ahlmark, Partners Chris Hardie and Rong Lu. Working out of studios located in Copenhagen, Aarhus, Shanghai and London, they are deeply commited to the Nordic architectural traditions based on democracy, welfare, aesthetics, light, sustainability and social responsibility. Received prestigious, national and international awards through the years including the Governor General's Medal in Architecture in Canada/ A+Award 2016/Årets Bygge 2016/Architectural Review MIPIM Future Project Awards 2015/World Green Design Product Award 2014/RIBA National Award 2013/RIAS Award 2013/ArchDaily Building of the Year Award 2011/LEAF Award 2011.

Nelson Mota

Graduated at the University of Coimbra, where he lectured from 2004 until 2009. Currently is an Assistant Professor at the TU Delft, in the Netherlands, where he concluded in 2014 his PhD with the title "An Archaeology of the Ordinary. Rethinking the Architecture of Dwelling from CIAM to Siza". Was the recipient of the Távora Prize(2006) and authored the book A Arquitectura do Quotidiano (2010). Is a member of the editorial board of the academic journal Footprint, and a founding partner of Comoco architects.

VAUMM

Is established in 2002 by Tomás Valenciano Tamayo, Jon Muniategiandikoetxea Markiegi, Javier Ubillos Pernaut, Marta Álvarez Pastor, and Iñigo García Odiaga(from left). Adding specialists in different disciplines, both within and outside the office, adding basic knowledge to provide solutions. Through this variable organization, the permanent office is enlarged thanks to multiple collaborations depending on the situation, focus on forming each project the best team possible. Recently, they received 1st prize at restricted competition for Dantzagunea-Arteleku-Scenic Arts and Dance & Cultural in Errenteria in 2013 and for the Remodelling of Molinao Park and Sports Center in Pasaia in 2012. Participated in the exibition entitled '20 Spanish offices in Mexico, 20 Mexican offices in Spain'.

Tom Van Malderen

His activities stretch from the traditional architectural practice to the field of architectural theory which he explores through writing, installations and lectures. After obtaining a master in Architecture at LUCA, Brussels(1997) he worked for Atelier Lucien Kroll in Belgium and in different positions at architecture projects, both in the UK and Malta. Lectured at the University of Aix-en-Province in France and the Canterbury University College of Creative Arts in the UK. Contributes to several magazines and publications, and sits on the board of the NGO Kinemastik for the promotion of short film.

XTU Architects

X stands for the unknown mathematical variable and TU for the suffix in situ-references to the geography and the landscape work done by Nicolas Desmazieres[right] and Anouk Legendre[left], who founded XTU in 2000. Has been developing a strongly innovative approach of cities and housing based on nature as a model. Among its many pioneering projects, often awarded at both architectural and environmental levels, are the French Pavillon at the Expo Milano 2015, the Museum of Wine in Bordeaux, and the JeonGok Prehistory Museum in South Korea.

©Marco Caselli Nirmal

Renzo Piano Building Workshop

While studying at Politecnico of Milan University, Renzo Piano worked in the office of Franco Albini. After graduation in 1964, started experimenting with light, mobile,temporary structures. Has collaborated with Richard Rogers from 1971 (Piano and Rogers), and with Peter Rice from 1977 (Atelier Piano & Rice). Renzo Piano Building Workshop was established in 1981 with 150 staff in Paris and Genoa. Has received numerous awards and recognitions including: the Royal Gold Medal, 1989; the Pritzker Architecture Prize, 1998; the Gold Medal AIA, 2008.

Loci Anima

Was created by Françoise Raynaud in 2002, Paris after 18 years at Jean Nouvel. The firm balances architecture, urbanism and innovation. Her spaces create links between people and nature, the human and the animal, the human and plant life in buildings designed to be living entities. English Partner architect, Jonathan Thornhill has worked for Jean Nouvel and led projects such as the Leuum museum in Seoul, and latterly for Shigeru Ban on the Centre Pompidou-Metz before joining loci anima in 2007. He has brought an added sensitivity to the detailed design of their projects.

164

Reiulf Ramstad Architects
Reiulf Ramstad was born in 1962 in Oslo, Norway. Studied at the University of Genoa and received Ph.D. from the University of Venice (IUAV). Founded his own office in 1995 and earned a reputation for creating bold, simple architecture with a strong connection to the Scandinavian context and impressive landscape in particular. Was nominated for the Mies van der Rohe Award in 2007, 2011, 2012, 2013, 2014 and won the WAN award for leading architects of the 21st Century in 2012. His main activities related to education and research at Norwegian School of Science and Technology (NTNU).

10

©Geordie Wood

OMA
Shohei Shigematsu[above] is Partner at OMA(Office for Metropolitan Architecture) and the Director of the New York office. Was born in 1973, Japan and graduated from the Department of Architecture and received M.Arch at the Division of Engineering, Graduate School of Kyushu University. After graduation, he has worked at the Toyo Ito Architects & Assosiats and NKS Architects. Studied at the Berlage Institute in Amsterdam before joining OMA in 1998. Is a design critic at the Harvard Graduate School of Design, where recently conducted a research studio entitled Alimentary Design, investigating the intersection of food, architecture and urbanism.

96

©Jaime Chinarro

Menis Arquitectos
Fernando Menis is a Spanish architect from the Santa Cruz de Tenerife, Canary Islands. Graduated from the Barcelona Institute of Architecture. In 2004, he creates Menis Arquitectos, an architectural studio based in both Tenerife and Madrid, as well as Valencia. Menis' designs are characterized by being sustainable and adaptable, representing low cost projects combining the natural elements of the urban landscape with architecture. Is constantly invited to seat as a Jury, to conduct workshops or as a Guest Lecturer at Harvard, Columbia NY, ESA Paris, TU Berlin, Akbild Wien. Currently is an Associated Professor at the University of Hong Kong (HKU), the European University of Canarias (UEC).

164

图书在版编目(CIP)数据

公共建筑 : 英汉对照 / 荷兰大都会建筑事务所等编; 孙探春等译. — 大连 : 大连理工大学出版社, 2017.6
(建筑立场系列丛书)
ISBN 978-7-5685-0808-7

Ⅰ. ①公… Ⅱ. ①荷… ②孙… Ⅲ. ①城市空间一建筑设计一汉、英 Ⅳ. ①TU984.11

中国版本图书馆CIP数据核字(2017)第120699号

出版发行:大连理工大学出版社
（地址:大连市软件园路80号　邮编:116023）
印　　刷:上海锦良印刷厂
幅面尺寸:225mm×300mm
印　　张:12.25
出版时间:2017年6月第1版
印刷时间:2017年6月第1次印刷
出 版 人:金英伟
统　　筹:房　磊
责任编辑:张昕焱
封面设计:王志峰
责任校对:张媛媛
书　　号:978-7-5685-0808-7
定　　价:258.00元

发　行:0411-84708842
传　真:0411-84701466
E-mail:12282980@qq.com
URL: http://dutp.dlut.edu.cn

本书如有印装质量问题，请与我社发行部联系更换。